Biotechnological Innovations in Food Processing

BOOKS IN THE BIOTOL SERIES

The Fabric of Cells
Infrastructure and Activities of Cells

Biomolecules - Extraction and Measurement
Biomolecules - Purification Strategies
Biomolecules - Analysis and Properties

Principles of Cell Energetics
Energy Resource Utilisation in Cells
Biosynthesis and the Integration of Cell Metabolism

Organisation of Genetic Information
Regulation of Gene Expression

Crop Physiology and Productivity

Functional Physiology
Defence Mechanisms

Bioprocess Technology: Modelling and Transport Phenomena
Operational Modes of Bioreactors

Microbial Growth and Cultivation
Plant Cell and Tissue Cultivation
Animal Cell Cultivation

Bioreactor Design and Product Yield
Product Recovery in Bioprocess Technology

Techniques for Engineering Genes
Strategies for Engineering Organisms

Technological Applications of Biocatalysts
Technological Applications of Immunochemicals

Biotechnological Innovations in Health Care

Biotechnological Innovations in Crop Improvement
Biotechnological Innovations in Animal Productivity

Biotechnological Innovations in Waste Treatment and Energy Resources

Biotechnological Innovations in Chemical Synthesis

Biotechnological Innovations in Food Processing

Biotechnology Source Book: Safety, Good Practice and Regulatory Affairs

BIOTECHNOLOGY BY OPEN LEARNING

Biotechnological Innovations in Food Processing

PUBLISHED ON BEHALF OF:

Open universiteit   and   **Thames Polytechnic**

Valkenburgerweg 167   Avery Hill Road
6401 DL Heerlen   Eltham, London SE9 2HB
Nederland   United Kingdom

Butterworth-Heinemann Ltd
Linacre House, Jordan Hill, Oxford OX2 8DP

OXFORD LONDON BOSTON
MUNICH NEW DELHI SINGAPORE SYDNEY
TOKYO TORONTO WELLINGTON

First published 1991

© Butterworth-Heinemann Ltd 1991

British Library Cataloguing in Publication Data
Biotechnological innovations in food processing.
– (BIOTOL series)
  I. BIOTOL II. Series
  664.0076

ISBN 0 7506 1513 3

Library of Congress Cataloguing in Publication Data
A catalogue record for this book is available
from the Library of Congress

Composition by Thames Polytechnic
Printed and bound in Great Britain by
Butler & Tanner Ltd, Frome and London

The Biotol Project

The BIOTOL team

OPEN UNIVERSITEIT, NETHERLANDS
Dr M. C. E. van Dam-Mieras
Professor W. H. de Jeu
Professor J. de Vries

THAMES POLYTECHNIC, UK
Professor B. R. Currell
Dr J. W. James
Dr C. K. Leach
Mr R. A. Patmore

This series of books has been developed through a collaboration between the Open universiteit of the Netherlands and Thames Polytechnic to provide a whole library of advanced level flexible learning materials including books, computer and video programmes. The series will be of particular value to those working in the chemical, pharmaceutical, health care, food and drinks, agriculture, and environmental, manufacturing and service industries. These industries will be increasingly faced with training problems as the use of biologically based techniques replaces or enhances chemical ones or indeed allows the development of products previously impossible.

The BIOTOL books may be studied privately, but specifically they provide a cost-effective major resource for in-house company training and are the basis for a wider range of courses (open, distance or traditional) from universities which, with practical and tutorial support, lead to recognised qualifications. There is a developing network of institutions throughout Europe to offer tutorial and practical support and courses based on BIOTOL both for those newly entering the field of biotechnology and for graduates looking for more advanced training. BIOTOL is for any one wishing to know about and use the principles and techniques of modern biotechnology whether they are technicians needing further education, new graduates wishing to extend their knowledge, mature staff faced with changing work or a new career, managers unfamiliar with the new technology or those returning to work after a career break.

Our learning texts, written in an informal and friendly style, embody the best characteristics of both open and distance learning to provide a flexible resource for individuals, training organisations, polytechnics and universities, and professional bodies. The content of each book has been carefully worked out between teachers and industry to lead students through a programme of work so that they may achieve clearly stated learning objectives. There are activities and exercises throughout the books, and self assessment questions that allow students to check their own progress and receive any necessary remedial help.

The books, within the series, are modular allowing students to select their own entry point depending on their knowledge and previous experience. These texts therefore remove the necessity for students to attend institution based lectures at specific times and places, bringing a new freedom to study their chosen subject at the time they need it and a pace and place to suit them. This same freedom is highly beneficial to industry since staff can receive training without spending significant periods away from the workplace attending lectures and courses, and without altering work patterns.

Contributors

AUTHORS

Mr J. A. M. van Balken, DSM Research, Geleen, The Netherlands.

Dr L. A. M. van den Broek, Agricultural University, Wageningen, The Netherlands.

Dr H. A. Busnach, Gist-brocades, Delft, The Netherlands.

Ir J. Bol, TNO-CIVO Food Analysis Institute, Zeist, The Netherlands.

Dr C. A. G. van Eekelen, Gist-brocades, Delft, The Netherlands.

Ir Th. G. E. Geurts, Gist-brocades, Delft, The Netherlands.

Dr Ing J. Kamphuis, DSM Research, Geleen, The Netherlands.

Dr J. Labout, Gist-brocades, Delft, The Netherlands.

Prof J. Maat, Unilever Research Laboratories, Vlaardingen, The Netherlands and Free University, Amsterdam, The Netherlands.

Dr P. J. van der Meer, Ministry of VROM, Leidschendam, The Netherlands.

Dr O. Misset, Gist-brocades, Delft, The Netherlands.

Dr B. Poldermans, Gist-brocades, Delft, The Netherlands.

Dr J. P. M. Sanders, Gist-brocades, Delft, The Netherlands.

Dr R. J. Siezen, Netherlands Institute for Dairy Research, Ede, The Netherlands.

Dr Ir J. Stadhouders, Netherlands Institute for Dairy Research, Ede, The Netherlands.

Dr D. A. Toet, Gist-brocades, Delft, The Netherlands.

Dr W. J. Quax, Gist-brocades, Delft, The Netherlands.

Prof A. G. J. Voragen, Agricultural University, Wageningen, The Netherlands.

Prof W. M. de Vos, Netherlands Institute for Dairy Research, Ede, The Netherlands and Agricultural University, Wageningen, The Netherlands.

Dr J. Wilms, Gist-brocades, Delft, The Netherlands.

EDITOR

Dr J. Green, Napier Polytechnic, Edinburgh, UK.

SCIENTIFIC AND COURSE ADVISORS

Dr M. C. E. van Dam-Mieras, Open universiteit, Heerlen, The Netherlands.

Dr C. K. Leach, Leicester Polytechnic, Leicester, UK.

Contents

How to use an open learning text

An open learning text presents to you a very carefully thought out programme of study to achieve stated learning objectives, just as a lecturer does. Rather than just listening to a lecture once, and trying to make notes at the same time, you can with a BIOTOL text study it at your own pace, go back over bits you are unsure about and study wherever you choose. Of great importance are the self assessment questions (SAQs) which challenge your understanding and progress and the responses which provide some help if you have had difficulty. These SAQs are carefully thought out to check that you are indeed achieving the set objectives and therefore are a very important part of your study. Every so often in the text you will find the symbol Π , our open door to learning, which indicates an activity for you to do. You will probably find that this participation is a great help to learning so it is important not to skip it.

Whilst you can, as a open learner, study where and when you want, do try to find a place where you can work without disturbance. Most students aim to study a certain number of hours each day or each weekend. If you decide to study for several hours at once, take short breaks of five to ten minutes regularly as it helps to maintain a higher level of overall concentration.

Before you begin a detailed reading of the text, familiarise yourself with the general layout of the material. Have a look at the contents of the various chapters and flip through the pages to get a general impression of the way the subject is dealt with. Forget the old taboo of not writing in books. There is room for your comments, notes and answers; use it and make the book your own personal study record for future revision and reference.

At intervals you will find a summary and list of objectives. The summary will emphasise the important points covered by the material that you have read and the objectives will give you a check list of the things you should then be able to achieve. There are notes in the left hand margin, to help orientate you and emphasise new and important messages.

BIOTOL will be used by universities, polytechnics and colleges as well as industrial training organisations and professional bodies. The texts will form a basis for flexible courses of all types leading to certificates, diplomas and degrees often through credit accumulation and transfer arrangements. In future there will be additional resources available including videos and computer based training programmes.

Preface

Within the last century, scientific and technological developments have changed natural fermentation processes into controlled and optimised processes capable of operation on a large (industrial) scale. Such developments have led to the growth of the food industry. We have, as we have adopted an urbanised way of life, become more and more dependent upon food which has been processed in some way: to preserve it; to enhance its flavour; to make it more convenient to prepare and/or cook.

In this text, we examine how recent biotechnological developments have led to improvements in food processing. The techniques of genetic engineering and protein engineering have given new impetus to biotechnological food conversions and processing. We examine how they are being applied to production of fermented foods such as cheese and yoghurt and products such as fruit juices.

The same developments have also given new impetus to the use of biotechnological products as food processing aids (ingredients added to improve the quality of processed food). We examine processes for the production of natural preservation systems, sweeteners, flavour enhancers and many other useful products of value to the food industry.

This text, written in a reader-centred learning style, is quite advanced. It assumes that the reader has knowledge of the structure, composition and properties of the major biologically produced chemicals, basic microbiology and the principles of genetic engineering. This text focuses on the application of these disciplines to food production and processing.

This is done in the form of a series of case studies each with its own special emphasis.

Substantial background material is included to provide the context and support needed to understand the processes and strategies under discussion. The text does not confine itself to technical matters but also considers the wider implications of applying biotechnology to food production. This includes issues relating to regulatory requirements and economics.

Scientific and Course Advisors: Dr M. C. E. van Dam-Mieras
Dr C. K. Leach

Acknowledgements

Grateful thanks are extended to the authors, editors and course advisors, and to all those who have contributed to the development and production of this book. They include Dr N. Chadwick, Mr R. I. James, Dr M. de Kok, Dr G. Lawrence, Miss J. Skelton, Professor R. Spier and Mrs M. Wyatt. The development of BIOTOL has been funded by COMETT, The European community Action programme for Education and Training for Technology, by the Open universiteit of The Netherlands and by Thames Polytechnic.

Project Manager: Dr J. W. James

Introduction

Introduction

1.1 The changing food industry

Biotechnology in food production has a very long tradition. It is believed that fermentative production of food (such as bread, beer, cheese and wine) has been practiced for more than 10 000 years. Ever since, mankind has tried to improve these processes. On the one hand these improvements consisted of a long evolution of many small changes that were triggered mainly by casual circumstances: the variability in raw materials; the availability of ingredients; the personal taste of the cook; etc. On the other hand, several major developments changed food production to a much larger extent. These major developments were:

- scale of production; from preparation just for the family, the food industry evolved through specialists such as bakers and butchers to multinational organisations engaged in food manufacture and distribution on a worldwide scale;

- from labour intensive to automated processes; a second large change came from the introduction of machinery that to a large extent replaced labour. Later, automation of this machinery further replaced the need for a large workforce. These trends are continuing to evolve;

- transport and globalisation, availability and taste; a third large change derived from the influence of travel and the transportation of goods. From our journeys, we learned to appreciate food from other parts of the world. We also gained independence from the seasons by the development of food preservation enabling us to enjoy 'seasonal' foods all the year round. Through the development of transportation, we now can obtain fresh foods from all over the world. In this respect a well developed packaging and conservation technology is essential.

1.2 Biotechnology and food manufacture

Biotechnology has been defined as the integrated use of biochemical and microbiological sciences and process technology in order to apply the possibilities and properties of micro-organisms, cultured tissues and cells and part thereof in a technological way (European Federation of Biotechnology).

processing aids

food
ingredients

fermentation

Taking this definition, we realise that our daily food contains many constituents that are produced by biotechnology. Many of these biotechnological products will be described in detail in later chapters of this text. We can discriminate between processing aids that help us to carry out different process steps, and food ingredients that will form a part of the final food product. Enzymes, catalysing reactions such as the conversion of starch to glucose or fructose that are used as sweaters, form a large group of processing aids. Apart from the processing aids and food ingredients, other biotechnological products are manufactured by fermentation processes. In addition to the traditional products (bread, beer, wine) we can identify enzymes, flavourants and additives as products of fermentation.

empirical approach

Historically food processing was based on an empirical approach. Traditionally, these improvements were introduced by trial and error. Later, more systematic approaches have been used. After the invention of the microscope and the early microbiological work of Pasteur, we began to understand the different processes that are responsible for the characteristics of a food.

Looking back to the evolution in food products and trying to predict future improvements, we indicate as the driving forces of change:

- trends in consumer demands;

- needs of the producer;

- governmental pressures.

1.2.1 Trends in consumer demands

In recent times, we can recognise several patterns in the choice of foods by consumers. We can particularly identify:

- a request for healthy and natural products;

- a desire for tasty products;

- an increasing use of convenience products;

- a growing selection of non-local products.

1.2.2 Needs of the producer

Economic and regulatory pressures on food producers result in the:

- greater automation of food processes;

- need to maintain flexible manufacturing;

- need to maintain Good Manufacturing Practice;

- continued awareness of the cost of the product.

1.2.3 Governmental pressures

Social and political pressures encourage governments to:

- introduce strict environmental regulation;

- change the conditions leading to the registration of food products;

- enforce strict legislation on production.

1.3 Major factors influencing trends in the food market

We identify cost, preservation, taste, consistency, colour, safety and health aspects as the major factors influencing changes in the manufacture and processing of foods.

1.3.1 Cost

cost of materials

changing from sugar to aspartame

A major factor affecting cost is the yield of product achieved per unit of raw material. The ability to switch from one material to another is also important. As an example: sugar (sucrose) has been used in all sorts of food manufacturing processes. In the seventies high-fructose corn syrups, produced by the enzymatic conversion of corn starch to a glucose and fructose mixture replaced sucrose to a large extent. Recently a large share of the sweetener market has been taken by aspartame, a biotechnological product based on a dipeptide which is two hundred times as sweet as sucrose on a weight for weight basis.

labour costs

yield

environmental concerns

cost of waste treatment

The cost of labour is another factor determining price. The efficiency of conversion of raw material to final product has always been an important issue but with the environmental concern the optimisation of yield has become more important. Governmental regulations often demand reduction in the production of waste. This may be a costly activity especially if much effort is needed to satisfy regulatory demands. Those companies that anticipate regulatory developments have time enough to re-model their processes in such a way that they can reduce their waste streams just by increasing the yield in their processes. Here, new biotechnological processes may be very important.

1.3.2 Preservation

fermentation technology and its importance in food preservation

The preservation of food has been the most important factor in the past. Seasonal crops or peak demand could be handled by good food preservation technology. Historically, fermentation with a desirable organism kept many infections out of the food. The fermentation of yoghurt and brewing beer, the fermentation of wine, the fermentation of vegetables and meat are all examples of biotechnological preservation techniques of the past. Now, many alternative methods of preservation are available. These include refrigeration, freezing, packing under sterile conditions in various materials, preservation chemicals etc. Contemporary biotechnology is contributing to improvements in the new as well as the traditional food preservation techniques.

1.3.3 Taste

taste development

Taste is principally a property of the raw food material. Again, historically, we learned how to vary taste by the ripening of the raw material and by processes like cooking, frying etc. In fermentation, taste is developed by the interaction of the food material and micro-organisms. We have also learnt to modify taste by the addition of spices. Since the perception of taste is still poorly understood, taste as a description of quality of food products is still approached largely in an empirical way.

1.3.4 Consistency

texture

appearance

mouth feel
additives

Like taste, consistency is a property of food that influences the perception of food products by the consumer. Consistency includes textural aspects, appearance and mouth feel and can be influenced by fermentation (bread), heating, cooling and acidification. It can also be influenced by the addition of all kinds of ingredients like thickening agents, emulsifiers etc.

1.3.5 Colour

Colour is also important. Although most food colours do not contribute to the taste of food products, the absence of the typical colour of a food product gives us the perception that the product is not fresh or does not have the proper taste. Modern biotechnology has great potential to help in the production of well-known natural and synthetic products. Currently, the application of some natural products is limited because of a shortage in supply. The use of some synthetic colours is restricted because they are regarded as potential hazards to health. Biotechnology offers opportunities to circumvent these problems.

1.3.6 Safety aspects

long distance

large numbers of consumers

demand

high product safety

Safety aspects become more and more important. Long logistic lines from site of production to consumption present potential risks from the outgrowth of all kind of pathogenic micro-organisms. The numbers of these would be limited and therefore would not cause any harm if the products were consumed immediately after production. This type of problem sets very high hygiene specifications for the food processing and packaging industry. It is paradoxical that, as our modern society suffers much less from infectious diseases than ever before, the standards at home are becoming more and more relaxed. It is therefore essential that the food producing industry needs to maintain very high product safety standards. This is especially so if we realise that a single food producing unit may supply food to many individuals over a large geographical area. The problems of food safety as well as their solution are essential elements of food biotechnology.

1.3.7 Health aspects

There is growing concern over how healthy the food we eat really is. Many consumers have become aware of the possible risks to health from eating a diet high in calories or animal fats. Biotechnology provides opportunities to reduce these risks. An example is the replacement of sugar by low calorie sweeteners.

1.4 The structure of the text

Food technology, of which microbiology, fermentation and biochemistry form an integral part, has been applied to optimise food production. The issues mentioned above (cost, preservation, taste, consistency and colour) are the basic 'rational' objectives for these improvements.

The remaining chapters examine the contribution of biotechnology to achieving these objectives. We do not simply confine ourselves to these objectives, but indicate some of the regulatory and economic issues that need to be considered in the application of biotechnology. This discussion is mainly based upon a case study approach. To put the case studies into context, we begin in Chapters 2 and 3, by giving an overview of the contribution of processing aids and fermentation in manufacture.

The challenge to food technologists is to recognise the potential of biotechnology to fulfil the food requirements of todays society.

Processing aids for the food industry: The state of the art

Processing aids for the food industry: The state of the art

2.1 Introduction

The food industry serves the function of supplying us with high quality, wholesome foods, all the year round and at a distance (in time and location) from the place of primary production (where the food was grown). To achieve this, good logistics of distribution and preservation methods such as refrigeration and deep freezing are employed. However, much of our food is processed in order to preserve it, make it more convenient to cook, to improve its organoleptic qualities, or to create alternative foods. In processing, use is frequently made of processing aids, which are specific ingredients which in some way help to preserve or improve the quality of the processed food. In this chapter we will provide an overview of how biotechnological products (micro-organisms themselves or products from them) are used as processing aids.

processing aids

2.2 Bakery products

Flour, yeast, water and salt are the basic ingredients of bread and other similar baked foods. For centuries it has been common to add other materials that have a positive effect on the handling of the dough and/or the quality of the baked product. A summary of the most common bread-improving ingredients is listed in Table 2.1.

∏ Which of the ingredients listed in Table 2.1 are involved in modifying a) the texture of bread and b) the speed of bread making? Are any of the ingredients listed in Table 2.1 added to bread to increase its food value?

Fats, emulsifiers and oxidising agents can affect bread texture. Fats, oxidising agents, reducing agents and soya flour can affect the speed of bread making. The ingredients are not really added to increase the food value, even though they might do so (for example soya flour).

Enzymes either from malt or micro-organisms play an important role and function in addition to the endogenous enzymes. Their functioning is best illustrated by describing their involvement in the various phases of bread making ie mixing, fermentation, baking and storage of bread.

Type	Function
Fat	reduce processing time, crumb softening, flavour
Emulsifiers:	
- Oxygenated	fat reduction, crumb softening
- Monoglycerides	anti-staling, crumb softening, volume
- Others (eg lecithin etc)	handling, tolerance, structure, volume
Oxidising agents (eg vitamin C, bromate)	reduce fermentation time, tolerance, structure, volume
Reducing agents (eg cysteine)	handling, reduce processing time
Sugars	fermentation rate, colour, flavour
Thickeners, gums	water retention, volume, yield
Fermentation dough concentrates	flavour
Soya flour	fermentation time, colour
Malt	various
Enzymes	various

Table 2.1 Bread-improving ingredients.

2.2.1 Mixing

mobility
resistance

starch

pentosans
arabinoxylan

gliadins
glutenins

Mixing starts with hydration of the constituents. When the flour particles are fully hydrated, the dough reaches its point of minimum mobility and maximum resistance. About one-third of the water is bound to starch which represents 65-70% of the flour. Approximately 5 to 10% of wheat starch will have been damaged during the milling process and it is that particular fraction which absorbs most of the water. The hemicellulose fraction (also called pentosans or arabinoxylan) constituting about 3% of wheat flour also absorbs about one-third of the dough water. The remaining one-third of the water is present in the flour protein fraction (10-14% of the flour). Of the protein fraction, 80-90% consists of the gluten proteins and comprises essentially two major groups; gliadins and glutenins, which are present in approximately equal amounts. The glutenins, being high molecular weight proteins are responsible for the elastic properties of the dough, while the low molecular weight gliadins are responsible for the viscous properties of the dough. During mixing and hydration, the protein particles that contain the gluten disintegrate and the glutenins and gliadins unfold. The high molecular weight (HMW) glutenins are partially broken down:

- by rupture of disulphide bridges through the formation of thiyl-radicals created by physical forces or chemical oxidation;

- by S-S interchange between the gluten proteins and the low molecular weight compounds like L-cysteine and glutathione;

- by hydrolysis by endogenous proteases.

2.2.2 Fermentation

elasticity

The gluten proteins (formed from glutenin breakdown) subsequently form a cohesive network through S-S bridge formation. This network entraps carbon dioxide formed during the fermentation stage and is also responsible for the elasticity which is so characteristic of wheat flour dough. Variation in amount and quality of the gluten proteins is therefore a determining factor in flour quality. Rye flour, for example, does not contain HMW glutenins and the major difference between European wheat and American wheat is in their gluten content (American wheat having the highest gluten content). Too high gluten contents may, however, lead to too 'bucky', non-elastic doughs, which require long mixing times. Reduction of these negative traits can be brought about by addition of protease preparations, mostly derived from *Aspergillus oryzae*.

∏ What is the function of yeast during the fermentation stage? (To answer this question, think of the products of aerobic or anaerobic metabolism).

The yeast converts carbohydrates in the flour to CO_2 (the end product of aerobic or anaerobic respiration). The CO_2, trapped in the network of gluten proteins, causes the dough to rise (leaven).

∏ Using what you have learnt about the function of glutens in bread making what do you think would be the effect of overdosing dough with protease enzyme?

Too much protease activity would cause the gluten proteins to be degraded extensively. This would cause the dough to lose its elasticity and would decrease its ability to retain CO_2 (and hence it would reduce leavening).

redox enzymes

The S-S bridge formation is an oxidative process. In baking, the various types of chemical oxidising agents used (see Table 2.1) can promote this. Various endogenous redox enzymes may also play a role (for instance peroxidases, ascorbate oxidase, dehydroascorbate reductase, glutathione oxidase and glutathione reductase). The similar effect of addition of redox enzymes such as glucose oxidase, polyphenol oxidase, peroxidase or ascorbate oxidase has been described but they are not as yet commercially used. Aggregation of gluten proteins can also be catalysed by lipoxygenase in the form of enzyme-active soya flour. The enzyme catalyses the oxidation of polyunsaturated fatty acids present in flour, to peroxides which can then play an oxidant role. In addition carotenoid pigments are destroyed in a co-oxidation process, and bleaching of the dough and product subsequently results.

sugar addition

α-amylase

Wheat and rye flour have only limited amounts of fermentable sugars. The concentration of 0.5% in mono- and disaccharides is insufficient to maintain yeast fermentation at a level necessary to obtain a satisfactory loaf volume. This can only be partly corrected by adding sugar, as the carbon dioxide production rate should coincide with the ability of the dough to retain gas. Sugar addition leads to rapid gas production and gas loss. Starch-degrading enzymes can provide the gradual formation of fermentable sugars by hydrolysing damaged starch. As the amounts of endogenous α-amylase in wheat or rye flour are very variable (depending on variety and weather conditions during growth), most flours are supplemented with this enzyme.

limit dextrins

β-amylase

gluco amylase

The α-amylase breaks down the starch polymers to smaller fragments called limit dextrins. These are further broken down to maltose by β-amylase (which is normally present in flour in sufficient quantities). Alternatively, fungal gluco amylase enzyme can be used. This enzyme splits off glucose molecules successively from the starch molecule, starting at the non-reducing end. These activities are discussed in more detail in Chapter 6.

∏ What do you think the effect of overdosing the dough with amylases would be?

The effect would be to rapidly produce sugar. The effect would thus be the same as the addition of too much sugar, that is too rapid gas production resulting in gas loss (and hence reduced leavening).

Genetic engineering has been applied to improving the uptake and use of sugars by bakers yeast. The genetically engineered strain has an improved transcriptional promoter and produces CO_2 more rapidly than the parent strain, reducing the time it takes for the dough to rise. Approval for the use of this organism in human food has been granted in The Netherlands.

2.2.3 Baking

When dough is well-risen in its final stage, it is brought into the oven to be baked. All enzymes continue to function during baking until the internal dough temperature reaches the inactivation temperature of the enzyme. For most fungal and cereal enzymes, this is around 65°C.

amylose

ovenspring

loaf volume

An important phenomenon during baking is the gelatinisation of starch. Gelatinisation is a process in which the starch granules swell by the uptake of water, lose their crystalline structure and exude the amylose (a constituent of starch consisting of about 400-800 glucose units in a linear chain). As a result of this process the concentrated starch suspension in the dough is transformed into a starch gel. The strength of this starch gel is determined by the amylose concentration. Recrystallisation of amylose during baking and cooling results in the formation of a rigid 3-dimensional network and this is thought to be responsible for the fixation of the crumb structure. A fast crumb setting, however, reduces 'ovenspring' resulting in a small loaf volume. Degradation of the gel-forming starch by α-amylase maintains the dough's flexibility for a longer period during ovening, allows greater expansion and produces a large loaf volume. The temperature of inactivation of cereal or fungal α-amylase (58-64°C) is lower than the gelatinising temperature of wheat starch, and so only a small part of the starch is degraded during ovening.

Excessive starch degradation is detrimental to bread quality because it reduces gel formation and leads to an extremely soft, gummy and open crumb structure. As bacterial α-amylase is heat-stable, dosing is very critical in order to prevent this over-degradation. Fungal α-amylase is less heat-stable and is preferred as it does not require such a narrow dosing range. Apart from higher loaf volumes, addition of fungal α-amylase leads to a softer crumb. This effect can be explained by:

• a lower crumb density as a result of higher loaf volume;

• the partial breakdown of starch leading to reduced firmness of the formed starch gel.

hemicellulose
and loaf volume

The insoluble hemicellulose fraction of flour depresses the loaf volume and gives a coarse crumb structure. Enzymes derived from the fungi *Trichoderma spp.* or *Aspergillus spp.* may be used for the degradation of hemicellulose. However as hemicellulose absorbs disproportionately more of the water (compared to protein or starch) its degradation can lead to release of bound water into the dough. This can result in sticky dough, which makes the bread difficult to slice.

2.2.4 Storage

hardening

retrogradation

shortening

After bread leaves the oven, various processes lead to deterioration of quality. The major problem is hardening (crumb firming), caused mainly by the recrystallisation (retrogradation) of starch during storage. This can be avoided by the use of shortening (fats) to produce a softer crumb structure or by limited degradation by α-amylase at low dosage.

Π What do you think would be the major limitation to the application of α-amylase to crumb softening? (Think about when you would have to add the α amylase)

The enzyme would have to be added to the dough prior to baking. During baking, the enzyme (or most of it) is likely to be inactivated by heat. This would eliminate α-amylase activity, or reduce it to negligible levels.

Although the role of industrially-applied enzymes such as α-amylase and proteases from microbial or cereal origin seems to be clear, detailed knowledge of the biochemical and biophysical events which take place during bread making is still lacking. Fundamental research on all these processes may allow us to identify new enzyme activities. These may help in overcoming the problems in the bakery area (ie flour variability, dough handling, dough stability and bread staling). Modern biotechnology will play an important role, especially by supplying pure enzymes without 'contaminating' side-activities.

SAQ 2.1	Answer true or false to each of the following.

1) α-Amylase can be usefully applied at the mixing, fermentation, baking and storage stage of bread manufacture.

2) Proteases, oxidases and gluco amylases can each improve leavening of bread.

3) α-Amylase leads to increased loaf volume during baking by increasing the concentration of amylose.

4) Protease is only applied to wheat flour with a high gluten content.

2.3 Fruit and vegetable juices

In addition to clear and cloudy fruit juices, the main industrial products today are concentrates. Since the manufacturer of fruit juices wants to deliver products with a constant quality all year round, and since the supply of fruit is seasonal, the harvested fruits are, after storage in cold warehouses, processed into stable concentrates. Usually the concentrates are sold to the actual juice manufacturers who dilute and pasteurise the juices before delivery to the customer.

Enzymes have found wide application in the production of fruit juices in a number of ways:

- for clarification of juices;

- to improve the pressing yield of the fruit;

- to overcome filtration problems;

- to increase the liquefaction rate of fruits;

- to improve the colour and flavour extraction;

- in the case of vegetable juice, to enable the total liquefaction of vegetable material.

The application of microbial enzymes to fruit processing is covered in detail in Chapter 7.

2.4 Dairy products

The use of enzymes for the processing of milk into cheese is one of the earliest examples in history of the application of enzymes in food technology. Thousands of years ago it was found that milk which was stored in a bag made of the stomach of a recently killed calf was converted into a semi-solid substance. Upon pressing, this substance produced a drier material which showed good keeping qualities. This process has evolved into the production of a wide variety of cheeses.

casein

The main constituents of milk are protein, sugar and fat (see Table 2.2). Casein makes up 80% of the protein material. Two types of casein can be distinguished: α- and β-casein which precipitate in the presence of Ca^{2+} ions, and κ-casein which is not sensitive to Ca^{2+} ions. Casein is found as stable micelles in milk. The κ-casein protects the micelles from Ca^{2+} induced precipitation. During the clotting or curdling of milk, the κ-casein is hydrolysed and loses its protective function. The micelles coagulate and

curd

precipitate, and the coagulated material (the curd) is separated from the soluble

whey

components (the whey). This is described in detail in Chapter 5.

Component	% in milk	% dry matter
Water	87.2	-
Protein	3.2	25
Fat	3.9	31
Lactose	4.6	36
Minerals	0.7	5.5
Others	0.4	2.5

Table 2.2 The main constituents of milk.

Apart from the proteolytic enzymes which are used to manufacture cheese, other enzymes which are active on milk sugar (lactose) or on milk fat, have been developed in the last two decades.

2.4.1 Cheese making

rennet

chymosin

rennin

pepsin

Traditionally, rennet [a mixture of chymosin (also called rennin) and pepsin] is used for the curdling of milk. Rennet is obtained by salt extraction of the stomachs of suckling calves. Only chymosin (EC 3.4.23.4) is capable of the specific hydrolysis of the κ-casein which results in curdling (coagulation) of the milk (see Chapter 5 for details). The ratio of pepsin to chymosin increases during aging of the calves. Pepsin, a protease, can have a negative effect on the flavour development of the cheese. For example, rennet containing more than 15% pepsin it is considered unsuitable for typical Dutch cheese manufacture.

Π Why do you think that younger calves should produce the greatest quantities of milk-clotting enzyme? (You may not know much about bovine physiology, but think of the diet of new-born calves and compare it with that of older animals).

Obviously new-born calves feed exclusively on milk. As the calves are weaned, the proportion of milk in the diet reduces. Thus the need for milk-clotting enzyme decreases with age. There is, however, a continued need for protein-hydrolysing enzymes (eg pepsin).

An increase in cheese consumption and, at the same time, a decrease in the amount of calves for slaughtering has caused a worldwide shortage of calf rennet. As a consequence, many different proteolytic enzymes have been investigated as alternative clotting enzymes. Although none of them exhibit the unique properties of chymosin, some alternative coagulants have been developed and have been applied to cheese manufacturing, especially for cheeses which require a short maturation time.

stability of
microbial rennet

The fermentation industry has developed alternatives based on micro-organisms. It turned out that proteases from *Mucor miehei* and *Mucor pusillus* show nearly the same specificity as chymosin. Only in cheeses which require a long maturation time can minor differences in flavour and texture be detected between calf rennet and microbial rennet. During the development of microbial rennets a common application problem was encountered: the thermostability of the microbial proteases. After pressing the curd, approximately 90% of the proteolytic activity is recovered in the whey. In the case of chymosin, after pasteurisation of the whey this activity is completely abolished. In the case of microbial proteases, some proteolytic activity is left after pasteurisation. This interferes with the formulation of the whey into diary products and baby foods.

This problem was solved by the enzyme manufacturers by mild oxidation of the microbial proteases: the thermostability of the enzymes was lowered considerably, while the initial activity was unchanged. Microbial rennets with reduced thermostability have been available on the market for ten years. The total sales of rennet in 1988 were estimated to be approximately $100 million, of which 30% was accounted for by microbial rennets.

DNA
technology

A completely new development has been the introduction of calf chymosin produced by micro-organisms such as yeast. Recombinant DNA technology was used to transfer the calf gene, carrying the genetic code for chymosin, into micro-organisms. Once the micro-organism was capable of producing the chymosin on a large scale, a new source of rennet was available to the market. The microbial chymosin is completely identical to the calf chymosin, it does not contain pepsin or other undesired enzyme activities, it is available from a virtually unlimited source and it will reduce the need to slaughter suckling calves. Such developments have been carried out in Europe and the USA, and are covered in detail in Chapter 5.

Π   Can you think of any disadvantages the use of genetically engineered chymosin in food might have? (See if you can write down a reason before reading on).

The major disadvantage that the use of such products in food has, is that the general public are afraid of possible harmful effects from eating it. However in this case the product (chymosin) is traditionally used in food, and as you will learn later, it is produced by micro-organisms also traditionally associated with food. Consequently it has been accepted by the industry and consumers.

2.4.2 The modification of whey

Whey is produced in large quantities as a by-product of cheese manufacture. Uses for it are as animal feed or as an ingredient of manufactured foods.

Lactose is the major constituent of milk and whey, where it accounts for 36 and 75% of the dry substance respectively. Lactose is a disaccharide (see Figure 2.1) composed of glucose and galactose. It has a low sweetness, it forms hard and sharp crystals at high concentrations and it causes digestibility problems for lactose-intolerant individuals. These properties make lactose, and hence whey, rather unsuitable for addition to foods. The problem can be overcome by hydrolysis of lactose into glucose and galactose. Acid hydrolysis of lactose produces unwanted by-products, so enzymatic hydrolysis is preferred.

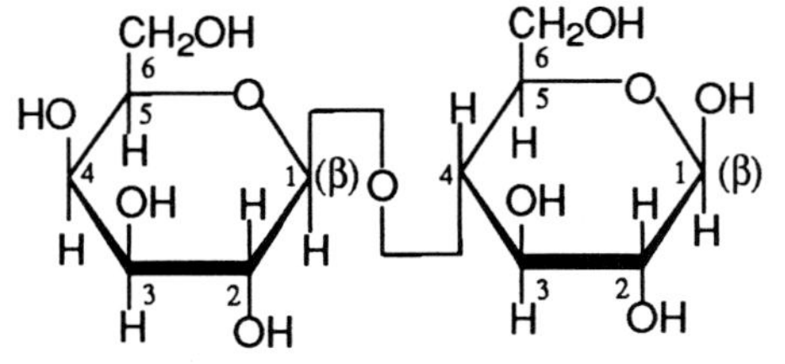

Figure 2.1 The structure of lactose.

β-Galactosidase   β-Galactosidase (E.C.3.2.1.23.) is produced by a variety of micro-organisms. Commercially available enzymes are usually derived from yeasts (such as *Kluyveromyces lactis*) or filamentous-fungi (such as *Aspergillus niger*). The major difference between the yeast and the filamentous-fungal enzyme is the pH optimum; yeast enzymes function in the range of pH 6-7, whereas filamentous-fungal enzymes show optimal activity at pH 4-5. For this reason yeast enzymes are usually employed for the hydrolysis of lactose in milk (pH 6.3-6.8) and filamentous fungal enzymes are applied in whey (pH 4-6). The major applications for β-galactosidase are found in:

- liquid milk and milk powder, to improve the product for lactose-intolerant individuals, or to increase the sweetness for milk-based drinks;

- concentrated milk products, to prevent crystallisation of sugars;

- fermented milk products, to increase the fermentation rate;

- whey as an animal-feed additive, to increase feed intake;

- whey as a food ingredient, to increase sweetness and prevent crystallisation (applied in ice-cream, confectionery and bakery products).

batch use of
β-galactosidase

The predominant mode of use of β-galactosidase is in a batch process. Since the optimal temperature of the enzyme is around 35°C, care must be taken to avoid microbial contamination. Milk is pasteurised prior to incubation with enzyme, the hydrolysis proceeds for 4 hours and before packaging or further treatment the milk is again pasteurised or sterilised by UHT (Ultra High Temperature) treatment. The hydrolysis can also be carried out at 6°C to retard microbial contamination. However, to produce the same degree of hydrolysis as that obtained after 4 hours incubation at 37°C, the incubation period must be extended to 20 hours. Alternatively UHT-treated milk can be used with the addition of sterile β-galactosidase, followed by aseptic packaging.

Since the enzyme is lost in all these batch processes, continuous systems using immobilised β-galactosidase have been developed. In such systems the enzyme is used repeatedly. Treatment of milk in this way has proved to be difficult due to contamination caused by continuous operation under moderate temperature and pH. Hydrolysis of whey is more successful since whey has a lower pH.

2.4.3 Enzyme-modified cheese (EMC)

flavour
production

Proteases and lipases from different micro-organisms are used for the production of cheese flavours. The technology consists of the incubation of a natural cheese slurry with specific proteases and lipases, usually at 37°C for a variable period (from 2 days up to 2 weeks), followed by inactivation of the enzyme and formulation of the end product into a paste.

∏ How do you think that protease and lipase enzymes in a cheese slurry would be inactivated? (By inhibitor, by pH adjustment, by heat, by oxidation?)

The answer is that heating is mostly used in foods to inactivate enzymes. The enzymes will be inactivated by high enough temperatures. The temperature and the heating time will affect the degree of inactivation.

Enzyme modified cheese is available in a variety of types (eg Cheddar, Parmesan and Edam) and with different strength with regard to the flavour concentration (eg 5 to 20 times the intensity of the normal cheese). EMCs are used as cheese substitutes in a number of products such as snacks, pizzas and cheese dips. Lipases are also used in the manufacture of certain Italian cheese varieties in order to give these cheeses their characteristic flavour. In this case, the enzyme is used to create a natural flavour, and not for the development of a flavour concentrate. The enzyme is added to the milk before the addition of rennin and is active under normal ripening conditions. An example is the use of Piccantase[R], a lipase produced by the fungus *Mucor miehei*, for the production of Romano and Provolone.

Likewise butterfat can be modified to give it a unique flavour. The modified fats are used in confectionary (toffees), as coffee whiteners and in bakery products.

∏ What do you think the most important properties of microbial enzymes used in dairy products would be? (Make a list before reading on).

Firstly the enyzme must be food grade, as food additives are strictly controlled: any enzyme added to foods must meet regulatory requirements (see Chapter 5 for more details). You may not have thought of this, but it is essential. More likely you suggested that the enzyme should be active at the appropriate pH and temperature. Also it should have the correct specificity of action during production and storage of the product.

2.5 The tenderization of meat

endogenous enzymes

Tenderness is a desirable quality of meat. Endogenous enzyme activity (neutral protease and collagenase) is the main factor contributing to the development of tenderness, ie to the conversion of muscle into meat. The biophysical factors which are involved in this process are not yet well understood. Natural maturation of meat takes place in approximately 10 days at 2°C. Slow maturation results in tender meat but has the disadvantage of moisture loss and shrinkage of the tissues. Attempts have been undertaken since 1940 to improve the tenderization process with exogenous enzymes.

exogenous enzymes

papain and bromelain

On a commercial scale, plant proteases such as papain (from papaya) and bromelain (from pineapples) are used. These proteases are capable of digesting connective tissue and muscle protein. A practical problem is how to achieve even distribution of the enzymes in the tissue. If preparations are sprinkled on the surface of the meat, the interior of the meat remains tough. Repeated injection under pressure into the meat is another possibility. Intravenous injection a few minutes before slaughtering has also been studied. In most countries treatment of meat with exogenous enzymes is controlled by legislation.

∏ What do you think would happen to the texture of your barbecue meat if you marinated it in fresh pineapple juice?

Since pineapple juice contains bromelain it would tenderize your meat. After a couple of hours you would even notice undesired softening and breakdown of the surface of the meat (another example of problems with enzyme overdosing).

∏ Meat is normally matured at refrigeration temperatures. Would you expect endogenous tenderizing enzymes to be highly active at such temperatures? (Think about the temperature at which these enzymes naturally work)

You would expect endogenous enzymes from warm-blooded animals to be most active at about 30-37°C, and this is the case. However meat cannot be matured at such temperatures because microbial spoilage would rapidly occur under such conditions. By comparison, many microbial proteases are active at lower temperatures than are mammalian enzymes.

2.6 The modification of oils and fats

triglycerides

The major components of oils and fats are triglycerides, and their physical properties depend upon the structure and distribution of their fatty acid groups. Natural oils and fats can be used directly in products either individually or as mixtures, but in many cases it is necessary to modify their properties, particularly their melting characteristics, to make them suitable for particular applications.

hydrogenation

inter-
esterification

Therefore, the oils and fats industry has developed processes which modify the composition of triglyceride mixtures. For example, fractional crystallisation is used to separate fats into solid and liquid fractions, and hydrogenation is used to reduce the unsaturation of fats (thereby raising their melting points). Chemical interesterification is used to change the physical properties of mixtures of fats by randomly redistributing fatty acid groups among the triglycerides.

advantages of
using enzymes

The potential advantages of using enzyme technology (compared to conventional chemical and physical procedures) are found in the specificity of enzyme catalysis and the mild reaction conditions under which enzymatic processes are operated. Enzyme catalysed reactions can be operated without the generation of by-products associated with the use of more severe chemical procedures, giving improved product yields and/or better product quality. The specificities of the enzymes can be exploited to generate products which are difficult to obtain by chemical procedures.

2.6.1 Selection of lipase

The enzymes used for modification of oils and fats are extracelluar microbial lipases. They are excreted by micro-organisms into the growth medium to catalyse the degradation of lipids, and can be produced on a large scale by fermentation.

lipase action

Lipases catalyse the hydrolysis of oils and fats to give diglycerides, monoglycerides, glycerol and free fatty acid. The reaction is reversible, and consequently microbial lipases also catalyse the formation of glycerides from glycerol and free fatty acid. Because of the reversibility of the lipase reaction, hydrolysis and re-synthesis of glycerides occurs when the enzymes are incubated with oils. This causes an exchange of fatty acid groups between triglyceride molecules giving interesterified products (Figure 2.2).

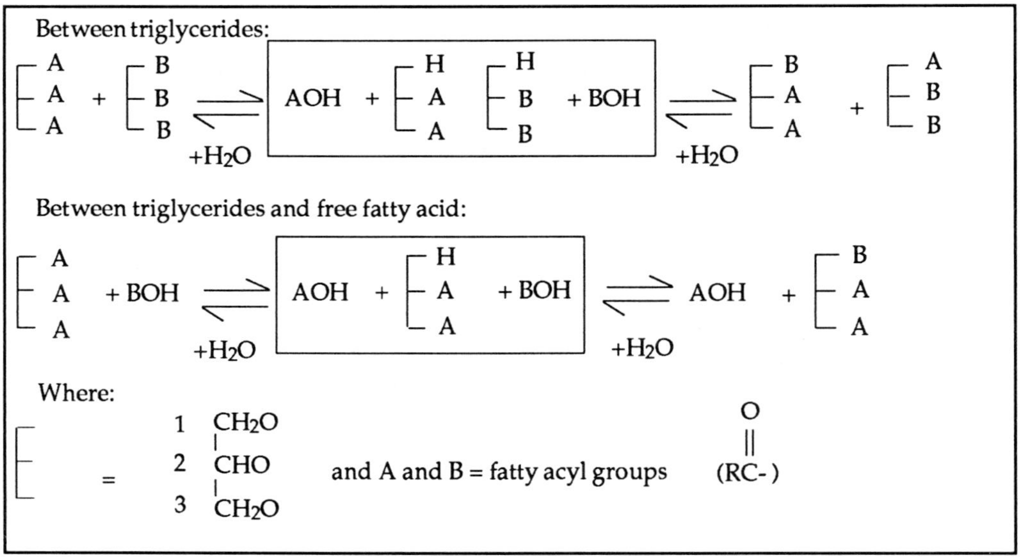

Figure 2.2 Mechanisms of lipase-catalysed interesterifications.

lipases active
in hydrophobic
environments

Microbial lipases are catalytically active in a predominantly organic environment containing very small amounts of water. Under these conditions the hydrolytic action of the enzymes is restricted by the limited availability of water, and high yields of interesterified triglycerides can be obtained. Mixtures of triglycerides and free fatty acids can be used as reactants for lipase catalysed interesterifications. In these cases free fatty acid exchanges with the fatty acyl groups of the triglycerides to form new triglycerides enriched in the added fatty acid.

The substrate specificities of lipases are crucial to their application as catalysts for modification of triglycerides. The enzymes can show specificity with respect to both the fatty acid and glycerol parts of triglycerides. However, most extracelluar microbial lipases are not highly specific with respect to the fatty acid groups found in the oils and fats used as raw materials for the edible fats industry, although reaction rates can vary with the chain length, and extent and position of unsaturation of the fatty acid group.

As regards the glycerol part of triglycerides, the specificities of lipases are of technical significance. Some microbial lipases are not specific and catalyse reactions at all three positions of glycerol. When lipases of this type are used as catalysts for interesterification of triglyceride mixtures, products containing a random distribution of fatty acid groups are obtained. These products have similar composition to those obtained by chemical interesterification.

regiospecificity

stereospecificity

A second group of lipases catalyse reactions specifically at both the outer (1- and 3-) positions of glycerol. These enzymes are said to show regiospecificity. This specificity can be exploited to produce triglycerides which are difficult to obtain by conventional chemical procedures. Regiospecific lipases catalysing reactions selectively at either the 1- or 3-positions of triglycerides would have very useful commercial applications. Microbial lipases show stereospecificity (specificity for stereoisomers) in reactions with many types of esters, but unfortunately stereospecificity in reactions with triglycerides has not been detected to date.

lipase stability

In addition to having the required specificity, lipases employed as catalysts for modification of triglycerides must be stable and active under the reaction conditions used. Lipases are usually attached to supports (ie they are immobilised). Catalyst activity and stability depend, therefore, not only on the lipase, but also the support used for its immobilisation. Interesterification reactions are generally run at temperatures up to 70°C with low water availability. Fortunately many immobilised lipases are active and resistant to heat inactivation under conditions of low water availability, but they can be susceptible to inactivation by minor components in oils and fats. If possible, lipases resistant to this type of poisoning should be selected for commercial operations.

Other factors to be taken into account when selecting an immobilised lipase are safety and acceptability. The enzyme used must be safe with respect to both operations in the factory and the consumption of the final product. The food product's acceptance by regulatory authorities will be assisted if the lipase is produced by a micro-organism which is already used for production of other enzymes used in food processing. This is the case with any enzyme used in food, as we discussed in Section 2.4.3.

2.6.2 Interesterification catalysts

reaction mixture

The reaction systems used for modification of triglycerides usually consist of a lipase catalyst and a small amount of water dispersed in a bulk organic phase containing the reactants and, if required, a water immiscible solvent. The small amount of water in the reaction system partitions between the catalyst and the bulk organic phase.

Lipases catalyse reactions at interfaces, and to obtain high rates of interesterification the reaction systems should have a large area of interface between the water immiscible reactant phase and the more hydrophilic phase which contains the lipase. This can be achieved by supporting the lipase on the surface of macroporous particles.

enzyme
adsorption
onto
macroporous
particles

Highly active catalysts have been produced by adsorption of lipases onto macroporous acrylate beads, polypropylene particles and phenol-formaldehyde weak anion exchange resins. Protein is bound, presumably essentially as a monolayer, within the pores of the particles. The large surface area of the particles ($10m^2g^{-1}$) means that substantial amounts of protein can be adsorbed, and the pores are of sufficient size to allow easy access of reactants to this adsorbed protein.

∏ When choosing a support for an immobilised enzyme, what other factors (apart from activity and access to the substrate) do you think need to be considered? (Think about the cost of producing the immobilised system and how it will be used).

As well as being active, the immobilised enzyme also needs to be stable (active for a long period) and the support must promote this. The support must also have appropriate mechanical characteristics: it should not disintegrate if used in a stirred tank reactor; it should produce even flow (without channelling) in a packed bed reactor. The cost of the support is also important.

2.6.3 Interesterification processes

Mixtures of triglycerides, triglycerides plus free fatty acids or triglycerides plus fatty acid alkyl esters are used as reactants in fat modification processes. These mixtures are exposed to lipases supported on macroporous particles in the presence of a small amount of water. Liquid substrates (oils) can be reacted without use of a solvent, but with solid reactants (fats) it is necessary to add a solvent to ensure that the reactants and products are completely dissolved in the organic phase. Various water immiscible solvents can be used, but hexane is preferred for commercial operation because this solvent is already used industrially for the processing of oils and fats.

The fat modification processes can be operated either in batches using stirred tank reactors or continuously with packed bed reactors.

∏ Which process do you think would be preferred: stirred tank reactors operated batch-wise, or packed bed reactors operated continuously? (Before reading on consider the likely overall yields of each type of operation and the cost of operation. Then make your decision).

productivity

Continuous reactors are likely to give the greatest overall specific productivity (quantity of product formed in a given time from a given quantity of enzyme), and therefore could be most cost-effective. In addition, in batch systems, the longer residence time involved can result in side reactions, leading to a decrease in the yield of triglycerides (the triglycerides are degraded). Continuous systems are also easier to monitor and to regulate by automation.

∏ Before reading on, see if you can draw a flow diagram for an interesterification process in which a triglyceride (eg palm oil) is to be incubated with stearic acid in the presence of a regiospecific lipase in order to increase the number of stearic acid residues in the 1- and 3- positions of the triglyceride. When you have done that compare it with the following description.

In a typical reaction, a feedstream consisting of refined palm oil and stearic acid dissolved in petroleum ether, is almost saturated with water (water content 0.06%), and then pumped through a bed of regiospecific lipase from *Mucor miehei* supported on diatomaceous earth. High catalyst activity as measured by an increase in the stearoyl content of the triglycerides can be obtained throughout 300 hours of continuous operation. Analysis of the triglyceride products shows that stearoyl groups are incorporated exclusively into the 1- and 3- positions, mostly in exchange for palmitoyl groups. This stearoyl incorporation results in the formation of 1(3)-palmitoyl-3(1)-stearoyl-2-oleoylglycerol (POSt) and 1,3-distearoyl-2-oleoylglycerol (StOSt). POSt and StOSt are the major triglycerides of cocoa butter, a valuable confectionery fat. Table 2.3 shows a comparison of the percentage content of palm oil mid-fraction before and after enzymatic modification, with natural cocoa butter.

Triglyceride	Palm oil mid-fraction (% dry weight)	Enzymically produced fat (% dry weight)	Cocoa butter (% dry weight)
StStSt	5	3	1
POP	58	16	16
POSt	13	39	41
StOSt	2	28.5	27
StLnSt	9	8	8
StOO	4	4	6
Others	2	1.5	1
StStSt	⌐stearate ⊦stearate ⌐stearate	POP	⌐palmitate ⊦oleate ⌐palmitate
POSt	⌐palmitate ⊦oleate ⌐stearate	StOSt	⌐stearate ⊦oleate ⌐stearate
StLnSt	⌐stearate ⊦linoleic ⌐stearate	StOO	⌐stearate ⊦oleate ⌐oleate

Table 2.3 Composition of cocoa butter equivalent prepared using enzyme technology.

With dry reactants, the rate of reaction is very low and no by-products are formed. The function of water is probably both to maintain the catalyst in a hydrated, active state and to generate essential reaction intermediates. For operation of commercial reactors the feedstream water content is selected to give the minimum level of by-product formation while maintaining an acceptable reaction rate.

2.6.4 Applications of lipase catalysed interesterification

The main current potential application of lipase catalysed fat modification processes is in the production of valuable confectionery fats. Chocolate contains approximately 30% cocoa butter, and this fat confers on chocolate its required crystallisation and melting characteristics. Cocoa butter is expensive, and the food industry has developed cheaper fat mixtures, known as cocoa butter equivalents, which can be used in place of cocoa butter in chocolate and related products. Cocoa butter equivalents are obtained by blending palm oil fractions containing high levels of POP (see Table 2.3 for nomenclature) with fractions from exotic tropical seed fats such as shea oil, illipe butter

cocoa butter

cocoa butter equivalents

and sal fat which contain high concentration of POSt and StOSt. Palm oil fractions are cheap and readily available, but the fractions derived from the exotic seeds are comparatively expensive and subject to variation in supply and quality. Therefore the oils and fats industry has developed alternative methods of obtaining cocoa butter equivalents by converting cheap palm oil fats to cocoa butter-like fats using enzymes.

use of inter-esterification products

Other reactants have been used for the production of cocoa butter equivalents by the enzyme technique. For example products enriched in StOSt and POSt can be produced by reaction of olive oil, high oleate safflower and sunflower oils, sal fats and shea oleine with stearic and palmitic acids or their esters. It is also possible to use lipases specific at 1,3 sites to produce triglyceride mixtures having useful functional properties in products such as margarines, low-calorie spreads and bakery fats. An example is the formation of triglycerides containing two long chain saturated fatty acid groups and one medium or short chain fatty acid group. These fats are effective hardeners for margarines and other spreads.

high value and low value products

It has been shown that lipase-catalysed reactions can be used for the large-scale production of modified triglycerides. At present the technology is being targeted to the production of comparatively high-value products such as confectionery fats. Wider application of the reactions to lower-value, higher-tonnage products will be dependent on the development of cheaper processes using more-productive and/or cheaper catalysts.

Fortunately there are indications that immobilised lipase catalysts will become more efficient and cheaper in the future. In the past, because of low fermentation yields, lipases have been expensive in comparison with the other main groups of extracelluar microbial enzymes such as proteases. Application of gene transfer technology to lipases could make them available at lower cost in the future. Considerable attention is also being given to the development of more effective supports for enzyme immobilisation.

better supports for enzyme immobilisation

A range of organic and inorganic materials are being investigated as potential enzyme supports, and parameters which affect the activity expressed by immobilised enzymes are being studied.

Our knowledge of the detailed structure of extracelluar microbial lipases and how they function as efficient catalysts for the hydrolysis and re-synthesis of triglycerides and other esters is scanty. However, the protein chemistry of the enzymes is now being studied in detail. This work will provide information helpful in the development of more effective catalyst systems, and may permit application of protein engineering techniques to produce modified lipases with altered specificities or improved stability.

⫪ What do you think the ideal properties of a lipase for use in fat interesterification would be? (Write down a list before reading on).

Ideal properties would be: food-grade; correct specificity of action (preferably 1,3 action on triglycerides); stable in anhydrous environments; active and stable at required temperature (and when immobilised); ease of immobilisation; low production cost.

SAQ 2.2	Which of the following favour the enzyme interesterification of oils and fats using enzymes rather than chemical processing?

1) Enzymes operate at relatively low temperatures.

2) Enzymes operating at low water contents show limited hydrolysis.

3) Enzymes can have specificities for 1 and 3 positions of triglycerides.

4) Enzymes can be non-specific for 1, 2 and 3 groups of triglycerides.

5) Enzymes cannot operate with dry (dehydrated) reactants.

6) Enzymes cannot operate on solid reactants.

2.7 Natural preservation systems

The previous sections have predominantly focused on the use of enzymes in food processing. In this section we turn our attention to the problems of food spoilage. Throughout history the problem of food spoilage has plagued man. Early attempts to preserve food centred on readily available substances and processes, such as using sugars, salts (lowering water activity, A_w), spices and wood-smoke. Today preservation also utilises such factors as: temperature (sterilisation, cooling, freezing); lowering water activity (drying); adjustment of pH; gases (CO, CO_2, ethylene oxide, propylene oxide, sulphur dioxide and ozone); organic acids (sorbic, acetic, benzoic, lactic, propionic acid); antibiotics; irradiation; packaging; various additives (formaldehyde, monochloroacetic acid, borates, nitrite, sulphite); various combinations of these factors. While these factors have ensured a constant supply of unspoiled food, there is a reaction against addition of chemical preservatives to food amongst many consumers. There is thus interest in developing preservation techniques which can be promoted as being 'natural'.

factors used in food preservation

Although the use of anti-microbial proteins and peptides is not widespread as yet, we are going to examine them here to emphasise their possibilities in natural preservation systems.

Enzymes/proteins can function as anti-microbials in several ways, such as:

- depriving spoilage organisms of an essential nutrient;

- generating substances toxic to spoilage organisms;

- attacking a cell wall/membrane component, thereby physically disrupting the cell or changing the permeability of the cell wall/membrane (ie microbicidal substances).

2.7.1 Depriving spoilage organisms of an essential nutrient

glucose oxidase-catalase

This is best illustrated by glucose oxidase-catalase which, in the presence of glucose, depletes oxygen when food is stored in a closed container. Thereby the growth of obligate aerobes is prevented.

These enzymes catalyse the reactions:

$$2 \text{ glucose} + 2O_2 \xrightarrow[\text{oxidase}]{\text{Glucose}} 2 \text{ glucono-6-lactone} + 2H_2O_2$$

$$2H_2O_2 \xrightarrow{\text{catalase}} 2H_2O + O_2$$

Other oxidases may work similarly. Catalase is only necessary if the oxidase generates H_2O_2 (hydrogen peroxide).

∏   Why do you think that hydrogen peroxide formation in food is undesirable? (Think about the chemical activity of hydrogen peroxide).

Hydrogen peroxide is a strong oxidising agent, and if formed in food can (like oxygen itself) cause unwanted oxidations, leading to browning and off-flavours. Besides the prevention of microbial growth by removal of O_2, the glucose oxidase catase system also prevents oxidation reactions. The enzymes are applied in fresh orange juice and mayonnaise dressings.

conalbumin

lactogerrin

avidin

Natural non-enzymatic proteins do exist in foods, which display analogous activity. Conalbumin is a protein in egg white (accounting for about 12% of the total egg white solids). This protein irreversibly binds iron ions, so making them unavailable for contaminating micro-organisms. This particularly effects Gram-positive bacteria such as *Micrococcus spp.* (which could otherwise grow and cause spoilage). Lactoferrin (lactotransferrin) is a protein in bovine milk which has similar activity. Avidin is also found in eggs, and binds biotin. Micro-organisms which have a strict requirement for this vitamin are therefore unable to grow in eggs.

2.7.2 Generating substances toxic to spoilage organisms

Oxidases

oxidases

Oxidases can generate H_2O_2 which, in a system devoid of catalase or peroxidase, is lethal to many micro-organisms. Xanthine oxidase, found in milk, is a natural example of such an enzyme. However, as we have seen, hydrogen peroxide in food can have unwanted effects.

Lactoperoxidase

SCNO⁻ ions

Lactoperoxidase coupled with a peroxide or a peroxide-generating oxidase, converts SCN^- to $SCNO^-$ (a very reactive and lethal ion for micro-organisms). SCN^- and lactoperoxidase are indigenous to milk, and so low levels of H_2O_2 added to milk serve as an effective preservative. Preservation of soft ice cream mix and pastry cream has been demonstrated using this method.

Myeloperoxidase

hypochlorous acid

halogenation

The myeloperoxidase-peroxide-halide (myeloperoxidase-H_2O_2-Cl) system has been suggested for food preservation based on the production of hypochlorous acid, and/or halogenation of microbial compounds to form chloramines. Widespread application is not only hampered by a cheap regular supply of the enzyme, but also by the bleaching reactions that occur concurrently in the food systems.

2.7.3 Microbicidal substances

lysozyme and bacterial cell disruption

Lysozymes (1,4 β-acetyl muramidases) cleave peptidoglycan (the major structural cell wall component of Gram-positive bacteria). Lysozymes from different sources differ in specificity of action, and spectrum of activity (range of organisms affected). The cell wall can be completely disrupted, leading to cell lysis. Gram-negative bacteria are unaffected, as peptidoglycan is a minor component of their cell walls and is protected by the lipopolysaccharide outer layer of the cell wall. If this layer is disrupted by agents such as EDTA, unsaturated fatty acids or polylysine, then the cells do become susceptible to lysozyme action.

Lysozyme derived from hen egg white has been applied commercially to prevent 'late blowing' of certain hard cheeses such as Grano, Brovalone, Emmenthal, Asiago and Monasio, by effectively preventing gas production by *Clostridium tyrobutyricum*, a common contaminant in cows milk. It has also been reported to be effective against the pathogens *Listeria monocytogenes* and *Clostridium botulinum*. Thermolability hampers a more widespread application, but the protein engineering work on stabilisation of lysozyme promises to overcome this problem.

∏ Why is the growth of *Listeria monocytogenes* and *Clostridium botulinum* undesirable in food? (The previous paragraph might give you a clue).

Both of these organisms can cause food poisoning. Thus it is important to prevent their growth in food.

chitinase and fungal cell wall disruption

Chitinases act on chitin, a structural component of fungal cell walls. The enzymes can act as a component of anti-fungal systems in food. The enzyme can be manufactured from the bacterium *Aeromonas hydrophilia*.

bacteriocins colicin nisin

Bacteriocins are proteins produced by bacteria which kill other bacteria. Some are so potent that one molecule entering a cell is enough to destroy that cell. Colicins produced by *Escherichia coli* have been well studied, but their very narrow host range limits their application. Nisin produced by group N *Streptococcus spp.* is currently used as a food preservation agent in many countries (in cheese, tomato juice, cream-style corn, chow-mein, meat slurries and beer). Nisin has been shown to be effective in inhibiting certain Gram-positive bacteria, but not Gram-negative bacteria, yeasts or filamentous fungi.

cecropins

maganins

bacteriocins

Bactericidal proteins have been identified in various higher eukaryotes and are induced upon infection. Cecropins are produced by insects, defensins and bacteriocins by mammalian neutrophils and maganins by vertebrates. The cecropins and maganins interact with the lipid bilayer of membranes. In the lipid membrane they probably form a type of ion channel, thus disturbing the ion balance of the micro-organism. Another explanation may be found in a surfactant action of the proteins. Their spectrum is rather broad, including several Gram-positive and Gram-negative bacteria, fungi and even certain enveloped viruses. Specific modifications (including fusion with the bee venom mellitin) have already shown that broadening of the host range is possible, without affecting higher eukaryotic cells. Modern biotechnological procedures will be important for the production of such proteins should their use in foods be adopted.

∏ It would be a useful form of revision to draw up a chart of devices used to 'naturally' preserve food together with the foods they protect.

The systems described above are unlikely to be effective preservation systems by themselves and should only be considered as an extension of currently used preservation methods. Together with hygienic processing procedures, milder preservation systems will ultimately lead to products that share palatability and higher acceptability with longer shelf life.

SAQ 2.3	Which of the following contribute to protection of eggs from spoilage by Gram-negative bacteria?

1) Presence of lytic enzymes.

2) Presence of immunoglobulins (antibodies).

3) Lack of available biotin.

4) Lack of available iron ions.

5) Lack of catalase.

6) Presence of bacteriocins.

SAQ 2.4	Answer true or false to each of the following.

1) Catalase increases the preservative effect of glucose oxidase by helping to remove oxygen.

2) Catalase increases the preservative effect of glucose oxidase by removing H_2O_2.

3) Catalase increases the preservative effect of lactoperoxidase by removing H_2O_2.

4) Catalase increases the preservative effect of xanthine oxidase by removing H_2O_2.

2.8 Flavours

sensory signature

Flavour is a very important component of the consumer's appreciation of a food, contributing to its smell and taste and interacting with the mouth feel and colour of the final product as schematically shown in Figure 2.3. The flavour tends to define foods and beverages in that it provides an individual sensory signature for the product.

∏ Is the property of sweetness confined to the presence of sugars in a food? (Use Figure 2.3 to help you answer this).

No it is not. Sweetness can be produced by some inorganic compounds and several kinds of organic compounds other than sugars.

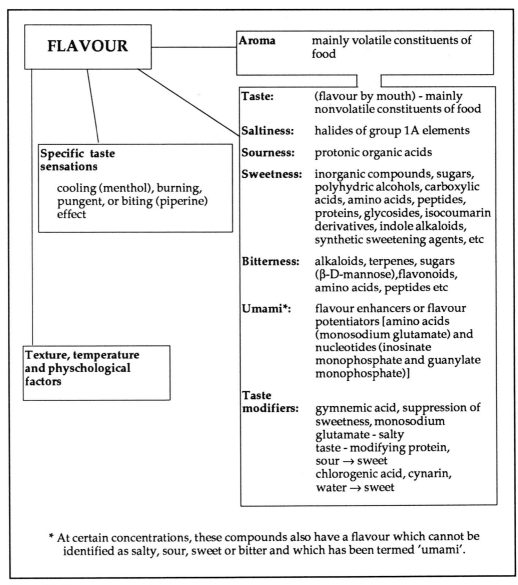

Figure 2.3 Flavour components.

Historically many food flavours have been generated by trial and error (very often when the primary concern was to enhance the stability of the food rather than improve the flavour). This has been the case for vinegar, cheese, yoghurt, beer and wine, in which the preservative effects of microbial generated molecules (such as acetic acid, methylketones, propionic acid and ethanol) were accompanied by desirable flavour development. The presence of complex mixtures of acids, alcohols, esters etc gives the individual character and identity to the food or beverage (see also Chapter 3). In addition, many flavours and fragrance raw materials for food have for several millennia been derived from plant sources (usually plants growing in countries in the botanically more diverse southern latitudes such as India, Indonesia and Africa). With the advent of sophisticated analytical procedures, complex flavour mixtures have been extensively

flavours from
plants

analysed and compounds classified into classes like green, sweet, fruity, roasted etc. Although various flavour compounds can be synthesised chemically and are identical to natural ones, there is a growing preference for 'natural' products by consumers.

This increasing interest in 'natural' products has placed more pressure on the production of natural flavours by extraction processes from plants and has also created interest in flavour production using biotechnological processes. These latter processes can be divided into two main groups, in which either cells or enzymes are used to generate flavouring complexes (multi-component flavour systems) or single flavour compounds.

2.8.1 Flavouring complexes

An example of a multi-component flavour system is cheese flavour. The use of specific strains of lactic acid bacteria and fungi to arrive at specific cheese flavour types is described in Chapters 3, 4 and 5. Another example is the use of short-chain fatty acid-specific lipases and proteases for the generation of Enzyme Modified Cheeses (EMC).

enzyme modified cheese

A blue cheese-type flavour is generated by *Penicillium spp.* (predominantly *P. roqueforti*). The free fatty acids that are generated from cheese by lipolysis are relatively toxic to *P. roqueforti*, and as a detoxifying mechanism the fungus converts these acids into methylketones (which are generally considered as the key flavour component of blue cheese).

methylketones

Methylketones arise as the result of two reactions, namely β-oxidation of fatty acids to β-keto acids followed by decarboxylation as shown below:

$$RCH_2CH_2COOH \quad \rightarrow \quad RCOCH_2COOH \quad \rightarrow \quad RCOCH_3 + CO_2$$

As both short chain- and long chain-fatty acids are partly metabolised, a variety of methylketones arise. The main ones that are produced when milk fat is used as a substrate are 2-pentanone, 2-heptanone and 2-nonanone.

2.8.2 Single flavour compounds

The other strategy is the biosynthesis, isolation and purification of individual flavour compounds. This approach involves exploiting bio-conversions such as oxidation and reduction, or synthesis by either microbial fermentation or by using specific enzyme systems.

∏ Before reading on, could you describe the difference between conversion (or transformation) and synthesis (or production)?

synthesis

conversion

transformation

precursor

When we say a micro-organism produces (or synthesises) a product, we mean that the product is formed as a normal metabolite of the organism from nutrients in the medium. Conversion (or transformation) by contrast, involves adding a compound (which is called the precursor) with a structure closely resembling the product to the culture (or microbial suspension). The micro-organism then converts the precursor to the desired product, usually by a simple step such as hydration/dehydration, oxidation/reduction and so on. This process is employed in the production of some steroids, antibiotics and amino acids.

Low molecular weight carbonyl compounds

diacetyl and
butter flavour

The characteristic buttery flavour of fermented dairy products such as sour cream or buttermilk is due to diacetyl, as described in Chapter 3. Lactococci produce diacetyl from citrate which is naturally present in milk. As can be seen from Figure 2.4 α-acetolactic acid is a key component in diacetyl biosynthesis and it has been proposed to increase the level of diacetyl-precursor by reducing the amount of α-acetolactate decarboxylase by genetic manipulation.

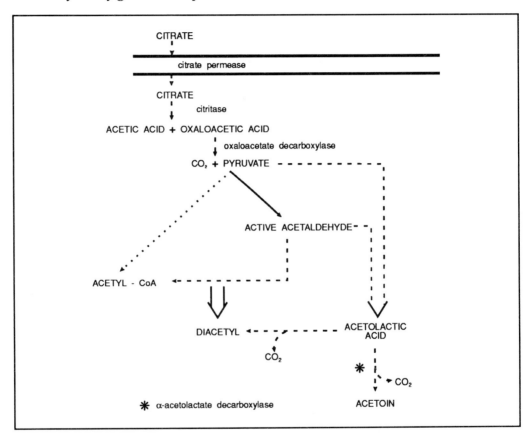

Figure 2.4 Diacetyl production from citrate.

Another important low molecular weight component is acetaldehyde, significant for the flavour of yoghurt and fruit. Procedures have been described for the conversion of ethanol to acetaldehyde by alcohol dehydrogenase (ADH) and alcohol oxidase (AOX).

Π   What do you think the major drawback of using enzymes such as ADH and AOX would be? You may get a clue from examining the reaction catalysed by ADH:

$$\text{ethanol} + NAD^+ \xrightarrow{\quad\text{alcohol dehydrogenase}\quad} \text{acetaldehyde} + NADH$$

Oxidoreductases require co-factors (such as $NAD^+/NADP^+$) in the reaction mixture. These cofactors are relatively expensive to produce, and so to operate effectively require a regeneration system. Such enzyme systems are thus more complex than those involving hydrolases (such as proteases, lipases and pectinases) which do not require cofactors.

regeneration of cofactors

A reaction scheme of ADH is given in Figure 2.5. A process with a conversion of 10-20% and a concentration of 2.5 gl^{-1} after a period of 9 hours has been reported. It is likely that modern biotechnological methods will greatly improve conversion rates. By linking ADH to catalase as shown, NAD^+ is regenerated allowing the reaction to take place in the presence of only small amounts of cofactors.

CH3CHO
NADH
FMN
H2O2
ADH
Light
O2
Catalase
CH3CH2OH
NAD+
FMNH2
O2
H2O

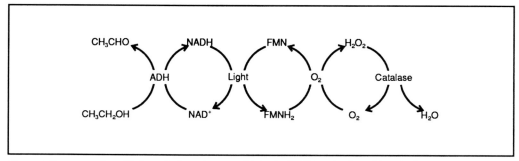

Figure 2.5 Enzymatic oxidation of ethanol to acetaldehyde (NAD^+ = nicotinamide adenine dinucleotide FMN = flavin mononucleotide).

Lactones

peach-like flavours

Lactones generally have very pleasant coconut- and peach-like fruity flavours and can be isolated from various classes of foods (fruits, vegetables, nuts, meat, milk products). Fermentative production has been described with the yeast *Candida spp.* using ricinoleic acid (12-hydroxy-octadecyl-9-enoic acid) as starting material (Figure 2.6). β-oxidation also plays an important role here. *Sporobolomyces odorus, Trichoderma viride* and *Polyporus porus* are other fungi which will carry out this reaction.

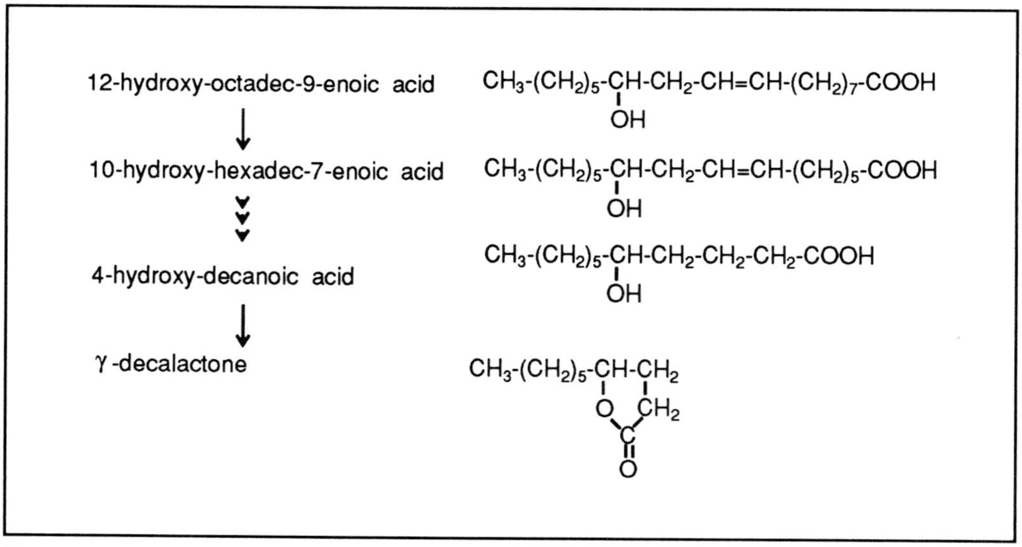

12-hydroxy-octadec-9-enoic acid
CH3-(CH2)5-CH(OH)-CH2-CH=CH-(CH2)7-COOH
10-hydroxy-hexadec-7-enoic acid
CH3-(CH2)5-CH(OH)-CH2-CH=CH-(CH2)5-COOH
4-hydroxy-decanoic acid
CH3-(CH2)5-CH(OH)-CH2-CH2-CH2-COOH
γ-decalactone
CH3-(CH2)5-CH-CH2
O CH2
C
O

Figure 2.6 Oxidative degradation of ricinoleic acid by *Candida lipolytica.*

Esters

fruity

Esters are responsible for fruity, flavoured aromas. Ethylacetate, ethylbutyrate, ethylisovalerate, ethylhexenoate are produced by various species of *Pseudomonas*, *Lactococcus* and *Lactobacillus* and yeasts such as *Hansenula anomala* and *Candida utilis*. A different strategy is the use of lipases for esterification, in which case conditions are similar as described for fat interesterification.

Pyrazines

roasted nutty
flavour

Pyrazines are heterocyclic, nitrogen-containing compounds which are often associated with roasted and nutty flavours. Methoxy alkylpyrazines have been found in a variety of vegetables including bell papers (specifically 2-methoxy-3-isobutylpyrazine), potatoes and green beans. Fermentative production of 2-methoxy-3-isobutylpyrazine has been described by *Lactococcus lactis*, *Pseudomonas perotens* and *P. taetrolens* and tetra-methyl pyrazine by *Bacillus subtilis* and *Corynebacterium glutamicum*.

Green components

cucumber,
apple and
tomato flavours

Green components like the C9 compound nonenal and nonadienal and the C6 compounds hexanal and hexenal play an important role in the overall flavour of cucumber and apple/tomato-type aromas. They arise through the degradation of unsaturated fatty acids via a lipoxygenase-catalysed formation of hydroperoxides followed by cleavage by a hydroperoxide lyase (Figure 2.7). Lipoxygenase from different sources differ considerably as regards their pH optima and substrate specificities. Thus soybean lipoxygenase produces mainly C13-hydroperoxides, whereas that from tomato produces C9-hydroperoxides. Enzymatic systems for the preparative-scale production of such green flavour compounds via the oxidation of fatty acids have not been described. The reason for this is the lack of availability of the lyase enzyme. Modern biotechnological procedures may thus have a great impact here.

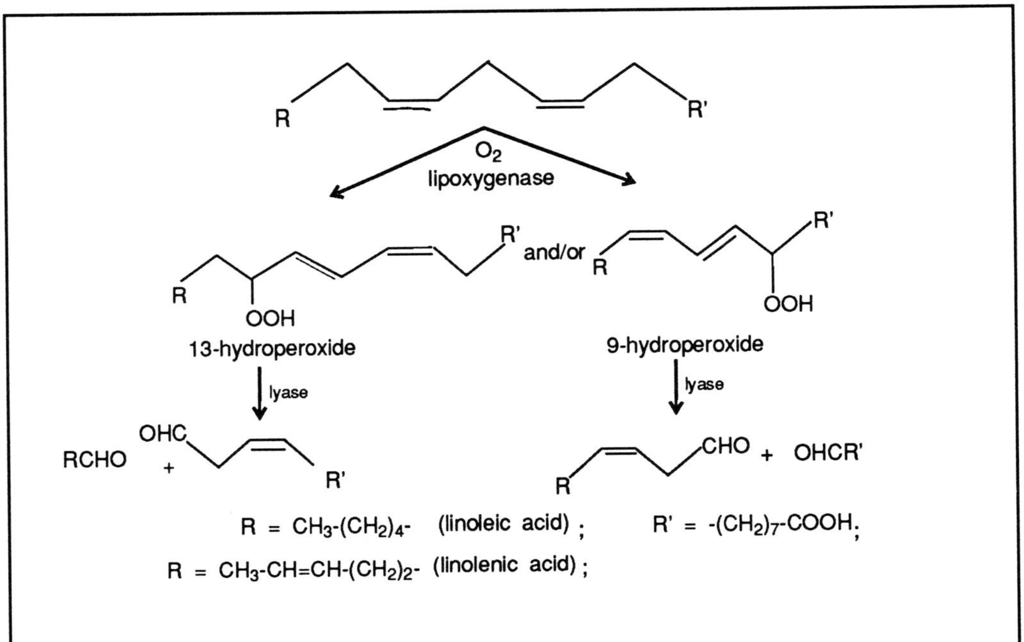

Figure 2.7 Enzymatic formation of aldehydes from unsaturated fatty acids.

Pungent tastes

isothiocyanates

Pungent tastes like those of mustard, cress and horseradish are due to the formation of isothiocyanates from odourless precursors known as glucosinolates by the action of the enzyme myrosinase (Figure 2.8). A system has been described for the use of immobilised myrosinase in the continuous production of horseradish aroma.

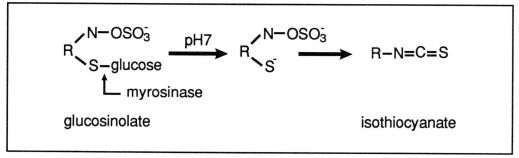

Figure 2.8 Action of myrosinase upon glucosinolates.

SAQ 2.5

In column A are listed various natural flavours. Match each with the appropriate compound listed in column B.

	A			B
i)	Butter		a)	2-Pentanone
ii)	Yoghurt		b)	Diacetyl
iii)	Roasted		c)	Acetaldehyde
iv)	Fruity		d)	α-Decalactone
v)	Peach		e)	Ethylacetate
vi)	Blue cheese		f)	2-Methoxyl-3-isobutyl pyrazine

SAQ 2.6

In column A are listed various natural flavours. In column B are listed various micro-organisms which produce natural flavours. From column B select that organism which itself produces the greatest number of the flavour compounds listed in column A.

	A			B
i)	Buttery		a)	*Trichoderma viride*
ii)	Yoghurt		b)	*Candida utilis*
iii)	Roasted		c)	*Corynebacterium glutamicum*
iv)	Fruity		d)	*Lactococcus lactis*
v)	Peach		e)	*Penicillium roqueforti*
vi)	Blue cheese			

2.8.3 Flavour enhancers

These materials are added to certain foods in order to enhance sensory responses. They can be produced by microbial fermentation or by enzymatic reactions, in combination with chemical methods. There are two nucleotide flavour enhancers: inosine 5'-monophosphate (5'-IMP) and guanosine 5'-monophosphate (5'-GMP) which can be produced by degradation of ribonucleic acid with 5'-phosphodiesterase. Microbial and plant sources of RNA have been described. Because of their higher RNA content of micro-organisms these provide a rich source of these compounds. RNA as a by-product of plant processing also provides a useful source. 5'-IMP arises through the conversion of adenosine 5'-monophosphate (5'-AMP) by adenosine deaminase. The production of 5'-IMP and 5'-GMP from the yeast *Candida utilis* is shown in Figure 2.9. The RNA used is a by-product of singe cell protein (SCP) production for food from this organism.

IMP
GMP

∏ The extraction and production of 5'-IMP and 5'-GMP is quite a complex procedure. Use Figure 2.9 to answer the following questions:

- How is RNA extracted from the cells?
- How is the extracted RNA hydrolysed?
- At what pH does RNA hydrolysis take place?
- How are AMP and GMP separated from CMP and UMP?
- What enzyme is used to convert AMP to IMP?

RNA is extracted from cells by hot alkali treatment. Extracted RNA is hydrolysed by fungal exonuclease. Hydrolysis takes place at pH 4.5.

AMP and GMP are removed from the mixture by adsorption on activated charcoal: CMP and UMP pass through the column and are discarded; AMP and GMP are adsorbed, then eluted in a methanol/ammonia mixture. Fungal adenylic acid deaminase converts AMP to IMP.

MSG

The monosodium salt of L-glutamic acid (MSG) is well known as a flavour enhancer and many hundreds of thousands of tons are used annually worldwide. MSG is produced by fermentation of sugars with *Corynebacterium glutamicum*. Strain selection, classical genetics and optimisation of fermentation have resulted in product levels of 30-50 gl^{-1}. Further improvement is expected by manipulation of the glutamic acid metabolic pathway through genetic engineering. The production of MSG is described in Chapter 8.

As well as the production of these separate components, there is a considerable production of crude mixtures through hydrolysis of plant, animal or microbial material. Yeast autolysates (yeast extracts) are widely produced and used as a flavour enhancer (especially savoury flavours).

2.8.4 Flavour precursors

importance of heat in flavour development

The method of processing food materials can have an important effect on the eventual flavour of the food. Heating in particular can give rise to reactions between various low molecular weight components like amino acids, peptides, sugars and lipids which lead to characteristic flavours in baked bread, boiled or baked meat etc. Very prominent is the Maillard reaction in which amino acids react with reducing sugars in a reaction known as the Strecker degradation pathway.

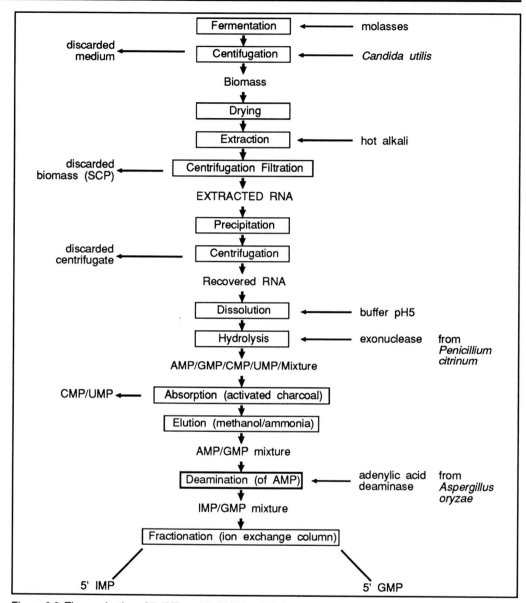

Figure 2.9 The production of 5'-IMP and 5'-GMP by RNA hydrolysis (simplified). AMP = adenosine 5'-monophosphate, GMP = guanosine 5'-monophosphate, CMP = cytosine 5'-monophosphate, UMP = uridine 5'-monophosphate, IMP = inosine 5'-monophosphate.

Biotechnology can also play a role in producing 'natural' flavour precursors. A typical example is the microbial production of 5-ketogluconic acid (5-KGA) by *Gluconobacter suboxidans* with glucose as substrate. The 5-KGA transforms upon heating into 4-hydroxy-5-methyl furanone, which is a major precursor material in meat flavour development.

2.8.5 Prospects

Currently it is estimated that the world market for food flavours is of the order of US $ 2.0 billion annually and is increasing at the rate of about 5 percent per year. Biotechnology will have an increasing impact on this speciality chemical industry.

Developments will be in the use of microbes and their enzymes in synthesis and conversion of flavour compounds such as we have outlined here. In this, gene manipulation and optimised extraction procedures for enzymes will play an important part. In addition to the application of micro-organisms to flavour production, development work is also underway to apply plant cell culture to this purpose. Such techniques have been developed for products such as capsaicin, the 'green pepper' flavour.

2.9 Biopolymers

thickeners
viscosifiers
emulsifiers
fillers

Biopolymers such as proteins and polysaccharides that are used as food ingredients generally have the property of easily dissolving or dispersing in water. These 'hydrocolloids' are often used as thickeners (gelling agents), viscosifiers, emulsifiers or fillers.

Commercially available proteins used as food additives are derived from both plants and animals. Plant sources are soya beans, cotton seed, sunflower seed, rapeseed and ground nuts (peanuts). They are often produced as by-products of oil extraction and are produced as flour (about 50% protein), as concentrates (about 65% protein) and as isolates (about 90% protein). Animal sources of protein are milk (casein and whey proteins), bones (gelatin), blood and eggs. Biotechnology may have applications in the improvement of such proteins by:

- increasing the level of essential amino acids (see Chapter 8);

- removing anti-nutritional factors (such as trypsin inhibitors);

- modifying the properties of the protein by proteases to allow easier processing (eg processing at lower temperature).

The commercially available polysaccharide-based hydrocolloids are derived from gums, mucilages, exudates and extracts of plant origin as listed in Table 2.4. They have varied functional properties owing to their structural diversity.

Type of polysaccharide	Example
Seaweed extracts	Agar-agar, carrageenan, Irish moss, alginate, laminarin, furcellaran
Lichen extracts	Icelandic moss
Plant extracts	Pectin, arabinogalactan (larch gum)
Plant exudates	Gum arabic, karaya gum, tragacanth gum, gum ghatti (Inidan gum)
Seed mucilages and gums	Locust bean gum, guar gum, tamarind mucilage
Plant starches	Cereals, potatoes, artichokes
Other natural hydrocolloids	Salep mannan

Table 2.4 Natural polysaccharides of plant origin.

∏ Read through the list in Table 2.4 and see if you can recognise the uses of some of these.

Starches are widely used as food thickeners in products such as gravy, custard and sauce mixes. Pectin is often added as a thickener to jam. Alginate or gums are often added as a thickener to sauces, relishes and spreads.

α-galacosidase in guar gum conversion

Recently a process has been developed to convert the rather cheap guar gum into a more valuable locust bean-like gum, by the use of the enzyme α-galactosidase. The enzyme, originating from the guar seeds ((*Cyamopsis tetragonoloba*), has been cloned and can be produced in large quantities using *Saccharomyces cerevisiae*. This enzyme has the peculiar property of being able to remove galactosyl residues that are α-1,4 linked to mannose residues in a polymannan backbone, at quite low water contents, which improves considerably the economy of the modification process. The galactose content of the galactomannan guar gum (38%) can thus be reduced to 10-15%, equivalent to that of galactomannan locust bean gum.

Bacterial polysaccharides represent a small fraction of the current biopolymer market, but have a large potential for the development of novel and/or improved products with the application of genetic engineering. Microbial polysaccharides with current or potential application in the food industry are summarised in Table 2.5. Many of these biopolymers find alternative uses. For example xanthan has been used in the formulation of cosmetics and dextrans have been used as plasma expanders (they increase the volume of plasma used for intravenous infusion.

Biopolymer	Organism (genus)	Composition and linkage	Molecular Weight (D)
Cellulose	*Acetobacter* *Agrobacterium* *Alcaligenes*	D-glucose β-1,4 linear	2.0×10^6
Dextran	*Lactobacillus* *Leuconostoc* *Streptococcus*	D-glucose α-1,6 α-1,3 branching	1.0×10^4-10^8
Alginate	*Azotobacter* *Pseudomonas*	D-mannuronate L-glucuronate β-10,4 linear	1.5×10^5
Curdlan	*Alcaligenes* *Agrobacterium*	D-glucose α-1,3 linear	-
Pullulan	*Aureobasidium*	D-glucose α-1,6, α-1,4 linear	2.5×10^5/ 1.0×10^7
Xanthan	*Xanthomonas*	D-glucose β-1,4 D-glucose- D-mannose- D-glucuronate β-1,4, β-1,2 α(1-3) branching	2.0×10^6/1.5×10^7
Chitosan	*Mucor*	D-glucosamine	1.7×10^4/ 1.3×10^5

Table 2.5 Microbial food-related biopolymers.

2.10 Sweeteners

The total consumption of sweeteners in the US has remained steady over the last 10 years at around 57 kg of sucrose equivalent per capita. In Europe this figure is slightly lower. The bulk of the sweeteners consumed in the 1960s was in the form of cane and beet sugar, sucrose (Figure 2.10). In the early 1970s however, enzyme technology opened up the way to a new class of sweeteners, based on starch instead of sucrose.

sucrose

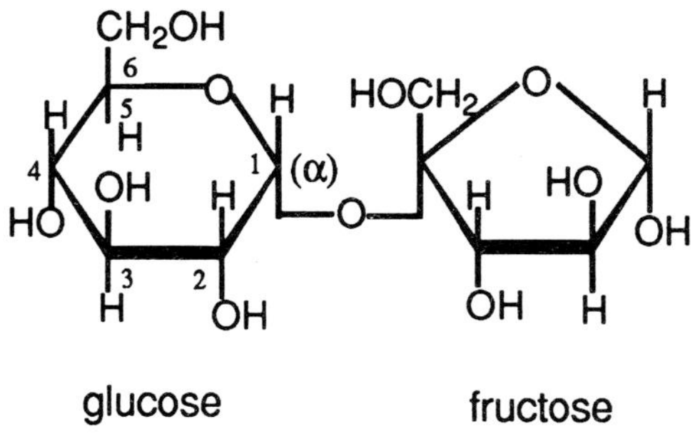

Figure 2.10 Structure of sucrose.

HFCS   High fructose corn syrup (HFCS) is a classic example of the impact of enzyme technology on the development of new products from existing raw materials. Soon after their introduction, these syrups exhibited an impressive growth at the expense of beet and cane sugar, and as a consequence sucrose consumption in 1985 had decreased to 31 kg per capita (Figure 2.11).

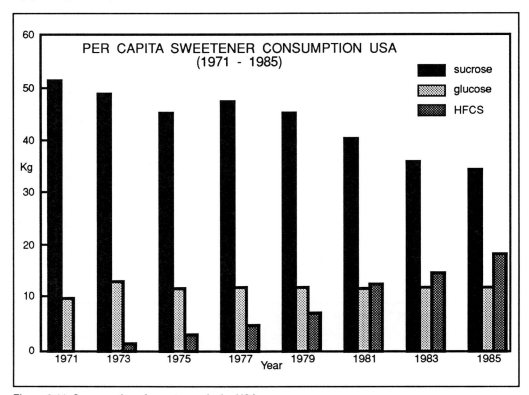

Figure 2.11 Consumption of sweeteners in the USA.

In Chapter 6 we will deal with HFCS development in more detail, so here we will confine ourselves to a general overview of the production of invert sugar (enzymatically hydrolysed sucrose) and the development of low-calorie sweeteners such as aspartame.

2.10.1 Invert Sugar

Cane and beet sugar can be converted to a 1 : 1 glucose : fructose mixture (called invert sugar), either by acid hydrolysis or by enzymatic hydrolysis (since sucrose is a disaccharide composed of one molecule of glucose and one molecule of fructose). The enzyme responsible for the hydrolysis of sucrose is called invertase. The virtue of invert sugar as a food ingredient compared to sucrose are:

invertase

• its higher solubility (and lower tendency to crystallise in concentrated solutions);

• its stability at low pH, and the enhancement of fruit flavours and higher sweetness due to the fructose present.

Before the development of HFCS, invert sugar accounted for 10% of sucrose used in the US and Europe. Since then the market has decreased, although there is still a steady, though small, market for the enzyme invertase with regard to special applications. A typical example is found in the production of confectionery food. The enzyme is injected into chocolate coated sucrose-containing candies, the interior of which is then transformed into a creamy consistency, characteristic of fondant-based confectionery. The enzyme is produced by bakers yeast, ie *Saccharomyces cerevisiae*, and is available in powder, liquid and immobilised formulations. The total market value is estimated to be around US $ 1 million per annum.

2.10.2 Alternative sweeteners

A large number of alternative sweeteners have been developed during the last 25 years. Lately the demand for low-calorie sweeteners has increased considerably, particularly in the soft drinks industry. A number of alternative sweeteners are shown in Table 2.6. Some of them find limited application, others are very successful, such as 55% HFCS and aspartame. Also the manufacturing processes vary considerably, for example thaumatin is a protein extracted from plants (see later), aspartame is a dipeptide synthesised by chemical or enzymatic methods and saccharin is chemically synthesised. Characteristics which determine the success of new sweeteners are the physical properties such as pH and temperature stability, textural aspects, the performance in relation to the price and the psychological effect on the consumer.

factors influencing the success of new sweeteners

Product	Relative sweetness
Sucrose	1.0
55% HFCS	1.4
Cyclamate	50
Aspartame	150
Saccharin	300
Thaumatin	3000

Table 2.6 Alternative sweeteners.

Aspartame

Aspartame, a dipeptide consisting of L-aspartic acid and the methylester of L-phenylalanine, is an example of an alternative sweetener which, since it was approved for use in soft drinks by the FDA (Food and Drug Administration, USA) in 1983, has gained a rapidly increasing market share. Originally aspartame was produced by a chemical process developed by Searle (USA); recently an enzymatic synthesis has been introduced. We will examine the production of aspartame in more detail in Chapter 8.

∏ Do you know what types of soft drink normally contain aspartame? (Here are some clues - To make a drink sweet, do we need to add as much aspartame as we would sucrose? If we are able to metabolise sucrose and aspartame would we derive the same amount of energy?)

Aspartame (trade name Nutra-sweet) is used as the sole (or part of the) sweetener in many low-calorie ('diet') soft drinks. You will find it in your favourite cola - whichever brand you prefer!

The sweet taste of aspartame depends on the L-conformation of the two amino acids, on the presence of the methylester and on the correct coupling of the amino acids.

Whereas α-aspartame is 150-200 times sweeter than sucrose, β-aspartame has a bitter taste (Figure 2.12).

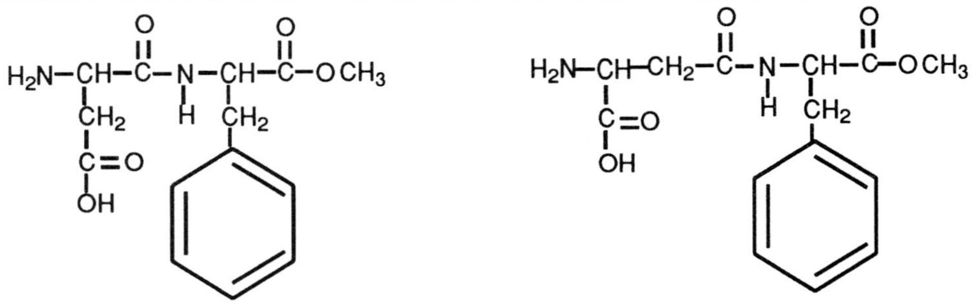

Figure 2.12 Structure of α- and β- aspartame.

To make aspartame, enzymic rather than chemical methods are used. To get the right product by chemical means, we would have to start the reaction with the amino acids in the correct stereospecific configuration, would have to block the β-carboxylic acid group of aspartic acid to stop it reacting and carry out the reaction at very cool (refrigerated) temperatures. The specificity of enzymes in catalysing the formation of aspartame from its constituent amino acids offers three advantages. These are:

- the use of ambient conditions instead of highly energy-intensive (refrigerated) systems;

- the use of racemic reactants which are cheaper than optically pure amino acids;

- there is no need to block the side-chain carboxylic group.

Thaumatin, the sweetest of them all!

On a weight basis, thaumatin, a protein from the berries of the plant *Thaumatococcus danielli*, is by far the sweetest compound known. Aqueous solutions of thaumatin range from 1300 times sweeter than sucrose, when compared at a 13% concentration, to about 5500 times sweeter when compared at a 0.6% concentration. Thaumatin is used to a considerable extent in Japan, and is marketed in the UK by Tate and Lyle. Several laboratories have undertaken research to develop a process for production of this protein by genetically engineered micro-organisms. Unilever (The Netherlands) and Beatrice Foods (USA) are the leading companies in this field.

2.11 Amino acids

applications of amino acid additives

In 1986 over 550 000 tons of amino acids were produced world-wide, representing a sales value of approximately US $1.5 billion. The major applications of amino acids and the respective market values are shown in Table 2.7. The application of amino acids in food can serve different purposes: as flavour enhancers, as seasonings, as nutritional additives and occasionally as improver ingredients (in bread).

Market	Value in US $ million
Food	500
Feed (food for animals)	600
Chemical	200
Pharmaceutical	130
Analytical	80

Table 2.7 Major markets for amino acids in 1986.

We will consider the production of amino acids in more detail in Chapter 8.

2.12 Organic acids

As with the production of amino acids, organic acids can be produced by chemical synthesis, by fermentation or by extraction from natural products. Enzymatic synthesis is not employed in the case of organic acids. Fermentative production is restricted to citric acid, gluconic acid, lactic acid and itaconic acid.

importance of citric acid

In volume, citric acid, which is exclusively produced by fermentation, is by far the most important organic acid, accounting world-wide for over 350 000 tons annually. Citric acid is generally produced by the fungus *Aspergillus niger* or the yeast *Candida lipolytica*, grown on molasses or other cheap carbohydrate sources. Yields on sugar up to 85% have been reported. Citric acid is used in food products to enhance the flavour, to prevent oxidation and browning, and as a preservative as a result of the pH lowering. It is also used as a raw material for chemical synthesis and for pharmaceutical purposes.

gluconic acid

Gluconic acid, which is also produced by fermentation with *Aspergillus* species, is only applied on a very limited scale in food.

lactic acid

Lactic acid has a world-wide production of approximately 50 000 tons. It is produced by fermentation (40%) and by chemical synthesis (60%). The latter results in a racemic mixture, whereas the former leads to the L-isomer. In nature, lactic acid is present as a racemic mixture. Lactic acid is used in food as an acidulant.

∏ Can you name another organic acid which is widely used as an acidulant in food? (Think of pickles!).

Acetic acid is widely used as an acidulant and flavouring compound in processed liquid foods such as pickles, sauces and ketchups. It is also used to pickle some vegetables and fish. It can be synthesised chemically, but much is produced by a 2-stage microbiological process. Stage 1 involves sugar fermentation by yeasts to produce malt beer, wine or cider. Stage 2 involves oxidation of the beer, wine or cider by the bacterium *Acetobacter spp.* to produce the corresponding vinegar. The vinegar itself can be used as a food ingredient, or the acetic acid recovered from it by distillation.

2.13 Vitamins

Vitamins are chemicals which are essential, in small amounts, for growth and development of living organisms. If they cannot be synthesised by the organisms, they must be taken up in the diet. Vitamins are used mainly as dietary supplements and in therapeutic applications. The exception is vitamin C (ascorbic acid) which is also used as a food ingredient. As a result of its relatively strong reducing power, it is used widely in food as an antioxidant and as an antimicrobial agent.

vitamin C
ascorbic acid

Π Can you think of a way in which vitamin C acts as an antimicrobial agent? (The clue is in its strong reducing ability).

Obligate aerobic bacteria (which may be associated with food spoilage) grow best is an environment in which there are chemically oxidised molecules. Such an environment is said to have a high redox potential (written as E_h or O.R. potential). Oxidising compounds such as nitrite can stimulate the growth of such bacteria. Reducing agents such as ascorbic acid can depress the growth of such bacteria.

Annually over 40 000 tons of vitamin C are produced for a bulk price of approximately US $9/kg. For the past 50 years ascorbic acid has been synthesised according to the seven-step Reichstein-Grussner process. In this process glucose is used as the starting material. Glucose is chemically reduced to sorbitol, which is converted to L-sorbose by the micro-organism *Acetobacter suboxydans*. This is followed by a number of chemical steps leading to L-ascorbic acid. Recently the production of vitamin C precursor (2-keto-L-gluconic acid) by *Escherichia coli* has been described through cloning and expression of the 2,5-diketo-D-gluconate reductase gene of *Corynebacterium spp.*

SAQ 2.7

In the table below are listed various processed foods or food products. In the appropriate columns, tick the products to which the enzymes might have been added.

Food/Food Product	Proteases	Amylases	Lipases
Bread			
Tenderized meat			
Cheese			
Interesterified oil			
Invert sugar			
Modified whey			
Modified protein thickener			
Organic acids			
EMC			

Summary and objectives

In this chapter we have examined the various ways in which biotechnological products can be used as processing aids in manufactured foods. Enzymes in particular are extremely useful in modifying the characteristics of food ingredients, so that the food has improved taste, texture and keeping qualities. Microbial products other than enzymes are used as flavours, sweeteners, thickeners and acidulants.

Now that you have completed this chapter you should be able to describe in outline:

* the roles of enzymes in preparation of bakery products;

* the roles of enzymes in the production of cheese and enzyme-modified cheese;

* the roles of enzymes in whey modifications

* the roles of enzymes in meat tenderization;

* the roles of enzymes in the interesterification of oils and fats;

* the roles of enzymes in natural food-preservation systems;

* the biotechnological production of flavours;

* the role of enzymes in the production of invert sugar;

* the biotechnological production of organic acids.

Fermented foods - an overview

Fermented foods - an overview

3.1 Introduction

In the fermentation industry, many micro-organisms are used for production of metabolites such as acids, alcohols, amino acids, enzymes and antibiotics. For food processing, fermentation products such as citric acid, acetic acid and glutamate are frequently used. For these fermentations, submerged culture in fermenters is aimed at production of the one particular product. The character of food fermentations is more *indigenous* complex. Vegetable or animal products are fermented by the indigenous mixed *mixed* microflora or by added starter cultures to improve shelf-life, nutritional value, flavour *microflora* and/or physical properties. The micro-organisms involved are therefore *starter cultures* multifunctional, and mostly form an integral part of the end product. Examples of fermented foods are sauerkraut, cheese, bread and many oriental foods such as soya sauce, miso and tempeh. Fermented foods have been produced throughout history. For instance, bread leavening and brewing by yeasts was practiced in ancient times in Egypt (ca 2500 BC). In China, fungi were intensively used in vegetable food preparation about 3000 years ago. The variety of fermented foods produced world-wide is enormous (several hundreds), with bacteria, filamentous fungi (moulds) and yeasts used as inocula.

preservation Renewed interest in food fermentations exists due to the fact that the preservation technology is relatively simple, and is regarded as being natural. Oriental food fermentations also offer good opportunities in the production of vegetarian meat flavourings and meat substitutes.

In this chapter we are going to examine several acid-based food fermentations and fungal-fermented oriental foods to provide an overview of the role of fermentation in *organoleptic* food processing. You will learn how microbial activity contributes significantly to food preservation, and desired organoleptic properties (taste, smell and texture) and leads to *conversion* food conversion (eg milk to cheese).

3.2 Acid-based food fermentations

endogenous In these types of food fermentations, animal and vegetable products are preserved by *spoilage* natural acidification due to formation of lactic acid from fermentable sugars (Figure 3.1) *organisms* by lactic acid bacteria belonging to the genera *Lactobacillus*, *Lactococcus*, *Streptococcus*, *Pediococcus* and *Leuconostoc*. In this book these organisms are abbreviated as: *Lactococcus* *endogenous* - L; *Lactobacillus* - Lb; *Leuconostoc* - Lc; *Streptococcus* - St; *Pediococcus* - Pd. The aim is to *and exogenous* produce microbiological as well as biochemical and chemical stability as a result of the *enzymes* fermentation process. This mainly involves the inhibition of the activity of endogenous spoilage organisms (those naturally associated with that food), although inhibition of *selective and* various endogenous enzymes (formed naturally from cells of the foodstuff itself) or *elective* exogenous enzymes (from external sources such as micro-organisms) is also important. *conditions* Each type of food thus requires its own conditions for fermentation, and the fermentation process has to account for changes in the organoleptic properties which are required. Normally the growth and activity of lactic acid bacteria is stimulated by the creation of selective and/or elective conditions, which do not promote the growth and activity of the endogenous organisms. This is achieved by using factors such as elevated fermentation temperature, anaerobic conditions (anaerobiosis) and/or reduced water activity (A_w) as a result of salt addition. These conditions stimulate the growth of lactic acid bacteria but do not promote the growth of many spoilage organisms.

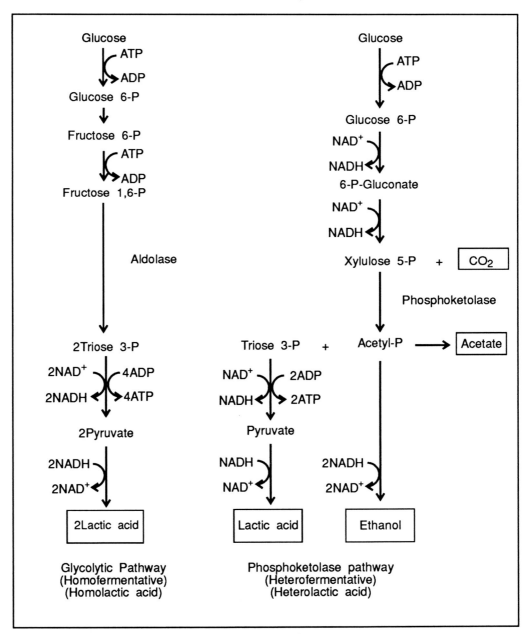

Figure 3.1 Outline of homofermentation and heterofermentation.

∏ What is the difference between water content and water activity? Can a food have a high water content but a low water activity?

Water content is how much water food contains (normally given as % w/w). Water activity is defined as the vapour pressure of water in the substance, divided by the vapour pressure of pure water, expressed as a fraction. The water activity is decreased by solutes, such as sodium chloride (salt) or sugars. In simple terms, high salt or sugar concentrations produce conditions in which cells cannot take up water. So although the water content might be relatively high, this water is unavailable for use by the cells if the A_w is low.

∏ Can you explain the difference between conditions which are selective and elective? (Have a go, before reading on).

inhibited

promoted

Selective conditions are those in which the growth of unwanted organisms is selectively inhibited by the presence of an inhibitory agent (such as salt or acid). Elective conditions are those in which the growth of the required organism is selectively promoted (for instance by providing them with a nutrient that they can use but the unwanted organism cannot).

Some unwanted yeasts and *Clostridium* species may however be active in an environment of reduced A_w and anaerobiosis. Additional acid formation by lactic acid bacteria resulting in pH lowering, produces further unfavourable conditions for these micro-organisms. Normally pH-values of 4.0-4.5 are needed to eliminate activity of clostridia. A combination of anaerobiosis, low pH, low A_w and a low concentration of fermentable sugars will together provide conditions in which few spoilage micro-organisms can grow, and will therefore produce a stable food.

∏ Which metabolic activities of micro-organisms do you think would lead to food spoilage? (Think about the ingredients of food and then about how micro-organisms might chemically change these ingredients).

The sugars in foods can be fermented to alcohols and acids. Proteolysis (protein degradation) can lead to the production of amines and sulphides. Lipolysis (lipid degradation) can lead to the production of fatty acids. These products cause changes in the taste, smell and texture of foods, possibly causing food spoilage. Such activity is, however, useful in the ripening of cheese (see Chapters 4 and 5).

starter culture

In the formation of pickles and sauerkraut, the selective conditions imposed by salt addition and anaerobiosis are generally enough to inhibit spoilage and allow endogenous lactic acid bacteria to predominate. Under certain conditions however some further degradation of lactic acid might occur through the action of micro-organisms. In the dairy and fermented meat industry inoculations of lactic acid bacteria (starter cultures) are employed.

homo-
fermentation

hetero-
fermentation

Lactic acid bacteria are said to be homofermentative (homolactic) if they produce lactic acid as the major (more than 90%) product of sugar fermentation. Heterofermentative (heterolactic acid) bacteria produce compounds other than lactic acid in greater proportions (Figure 3.1). In the next section we will be discussing lactic acid production. In some cases, we will refer to lactate. Remember that lactate is the ionised form of lactic acid. Thus:

$$CH_3CHOHCOOH \; \rightleftarrows \; CH_3CHOHCOO^- + H^+$$
$$\text{lactic acid} \qquad\qquad \text{lactate}$$

∏ What compounds other than lactic acid are end products of heterolactic acid fermentative organisms?

CO_2, acetate and ethanol are the other products of heterolactic acid fermentation.

3.3 Acid-fermented dairy food products

3.3.1 Introduction

range of
substrates

inocula

process
conditions

A variety of lactic acid bacteria and other micro-organisms are used in the production of a wide range of fermented dairy products. In Europe many products and cheese varieties are based on milk fermentations. This partly reflects the large cultural differences that exist between regions in this part of the world. However, the observed range may also be attributed to the following variables. Firstly, the milk used for the fermentations may be derived from cows, sheep, goats or other mammals, and is either used directly or after treatments such as pasteurisation, lactose hydrolysis or ultrafiltration. Secondly, the fermentation may be either spontaneous, or initiated by the addition of specific starter cultures or material from previous fermentations or even combinations of those practices, and may proceed at different temperatures for various times. Thirdly, the fermented milk is not necessarily the final product since it may be processed to a more concentrated product with reduced water activity, or blended with other ingredients. During and after the fermentation and processing stages, various additions may be made such as enzymes (including coagulants), salt, fruits, herbs, spices, sugars and natural colouring agents in order to increase or change properties such as flavour, preservation and appearance. We are going to consider only the main fermented dairy products that are made on an industrial scale from cow's milk, namely cheese, yoghurt, buttermilk and butter.

The main micro-organisms used in the production of products from cows milk are given in (Table 3.1).

∏ Which genera of bacteria are mainly used in the production of fermented dairy products? Are they all lactic acid producers? (You may get a clue from reading the names of the organisms in Table 3.1).

Lactococcus spp. are used in the production of most fermented dairy products. *Leuconostoc spp.* are used in many. These organisms are lactic acid producers.

The complexity of the microflora reflects the variety that exist in the products. The main process that contributes to the production of fermented dairy products is the fermentative conversion of the milk sugar, lactose, into lactic acid and, in some cases, other metabolites. This has the combined effect of eliminating or reducing the sugar content of the milk and decreasing the pH of the final product.

Product	Main lactic acid bacteria								Other micro-organisms
	L.lactis subsp lactis	*L.lactis subsp. cremoris*	*L.lactis subsp. lactis var. diacetylactis*	*Leuconostoc spp.*	*St. thermophilus*	*Lb. bulgaricus*	*Lb. helveticus*	*Lb. acidophilus*	
Buttermilk	X	X	X	X					
Fermented Butter (traditional)	X	X	X	X					
Fermented Butter (recombined)	X	X	X	X			X		
Yoghurt					X	X			
Bio-yoghurt					X	X		X	
Acidophilus milk	X	X	X	X				X	
Biogarde					X			X	*Bifido-bacterium*
Kefir	X	X							*Lb casei* *Lb caucasicus* *Saccharo-myces spp.*
Soft cheese:									
Quark Cottage	X	X	X	X					
Brie	X	X	[X]	[X]					*Penicillium camemberti/candidum*
Camembert	X	X	[X]	[X]					*Penicillium camemberti/candidum* *Coryne-bacterium spp.*
Kernhem	X	X	[X]	[X]					*Coryne-bacterium spp.*
Semi-soft Cheese:									
Munster	X	X							*Coryne-bacterium*
Stilton	X	X		X					*Penicillium roqueforti*
Hard Cheese									
Cheddar	X	X							
Gouda-Edam	X	X	X	X					
Emmenthaler Gruyere					X	X	X		*Propioni-bacterium spp.*
Maasdam	X	X	X	X			X		*Propioni-bacterium spp.*
Parmezan	X	X			X	X			
Mozzarella	X				X	X			

Table 3.1 Species involved in the production of fermented dairy products.

Note that the nomenclature of lactic acid bacteria has been revised recently. *L. lactis* subsp. *lactis* and *L. lactis* subsp. *cremoris* were formerly designated *Streptococcus lactis* and *Streptococcus cremoris*, respectively. In addition, the classification of the yoghurt bacteria *Lb. bulgaricus* and *St. thermophilus* has been changed into *Lb. delbrueckii* subsp *bulgaricus* and *St. salivarius* subsp. *thermophilus*, respectively. However, since in various countries yoghurt is defined by the law as a fermented milk product containing *Lb. bulgaricus* and *St. thermophilus*, the former designations have been used here. In addition, the new taxonomy has been proposed on the basis of phenotypic characteristics including biochemical and physiological properties. At present, however, it has been generally accepted that genotypic markers such as the primary sequence of rRNAs are the ideal candidates to classify bacteria and to propose new nomenclatures. This has recently been carried out on 16S rRNAs for species of *Leuconostoc*, *Lactococcus* and *Lactobacillus*. It is expected that extended use of these molecular markers will greatly help taxonomists, and provide a sound basis for strain identification for producers and legislators.

∏ It might be of help to you to remember these names by writing out a summary sheet. Use the headings:

Previous name	New name	Use

You will be able to add to this list as you read through this chapter.

∏ Would a decrease in sugar content assist in the preservation of milk products?

Yes. Milk is a highly unstable food, as it forms such a suitable growth medium for many micro-organisms. If the sugars are partly, or completely used up, then spoilage organisms will be less able to use the milk as a growth medium.

The lactic acid bacteria which are the main producers of acid are the mesophilic *Lactococcus* and *Leuconostoc* species that have optimal growth temperatures between 20 and 30 °C and the thermophilic *Lactobacillus* and *Streptococcus* species that are used at temperatures up to 45 °C. Apart from the production of lactic acid, these bacteria have other functions, including the formation of aromas and in some cases the production of extracellular polymers that contribute to the viscosity of the product. Other micro-organisms that contribute to producing of fermented dairy products are listed in Table 3.1. They include bacteria, yeasts and moulds (filamentous fungi) which in some cases are designated as secondary microflora. These micro-organisms usually contribute to development of flavour and other organoleptic properties and may also be involved in the final appearance of the product.

∏ Where lactic acid production is more important than flavour development, homofermentative lactic acid bacteria are used, rather than heterofermentative ones. Why do you think this is? (Examine Figure 3.1 for a clue).

Homolactic organisms will produce a higher yield of acid from sugar (that is the quantity of acid produced from a particular quantity of sugar). This is because lactic acid is the main end product of sugar utilisation. Heterolactic organisms convert some of the sugar to products other than lactic acid.

We will now examine the production of fermented dairy products, and see what roles lactic acid bacteria have in their production.

3.3.2 Activity of dairy starters and secondary microflora

The most important metabolic systems in lactic acid bacteria and other micro-organisms used in milk fermentations, relate to the presence in cows' milk of:

- lactose (approximately 44g l^{-1}), which is the source of fermentable sugar;

- citrate which can be metabolised by certain lactic acid bacteria;

- a high content of milk protein (approximately 32g l^{-1});

- a high content of milk fat (up to 40g l^{-1}).

Conversion of these compounds results in the production of metabolites that are characteristic of fermented dairy products.

Lactose metabolism

In all milk fermentations it is essential that the milk sugar lactose is converted rapidly into lactic acid. As may be expected, efficient mechanisms for the transport (into the cell) and degradation of lactose are found in the mesophilic and thermophilic lactic acid bacteria used as starter cultures. These include *L. lactis*, *Leuconostoc*, *St. thermophilus*, *Lb. bulgaricus* and *Lb helveticus*.

PTS   Two basically different mechanisms for transport and hydrolysis of the disaccharide lactose are found in these lactic acid bacteria and these are shown in Figure 3.2. In all industrially-used strains of *L. lactis*, lactose is phosphorylated during transport by a phosphoenolpyruvate (PEP) - dependent phosphotransferase system (PTS), which is mediated by an enzyme called II-lac. The intracellular lactose 6-P is split by a phospho-β-galactosidase to glucose and galactose 6-P.

ΙΙ   What is the net gain in ATP from uptake and utilisation of one molecule (mol) of lactose in *L. lactis*? (Calculate this using Figure 3.2 before reading on).

- The conversion of glucose (1 mol) to glyceraldehyde 3-P (2 mols) uses 2 ATP mols.

- The conversion of galactose 6-P (1 mol) to glyceraldehyde 3-P (2 mols) uses 1 ATP mol.

- The conversion of 4 mols glyceraldehyde to 4 mols PEP produces 4 ATP mols.

- Assuming 1 PEP mol is used in the transferase system, the conversion of the remaining 3 glyceraldehyde mols to lactate produces 3 ATP mols.

Thus 3 ATP mols are used and 7 are produced, resulting in a gain of 4 ATP mols.

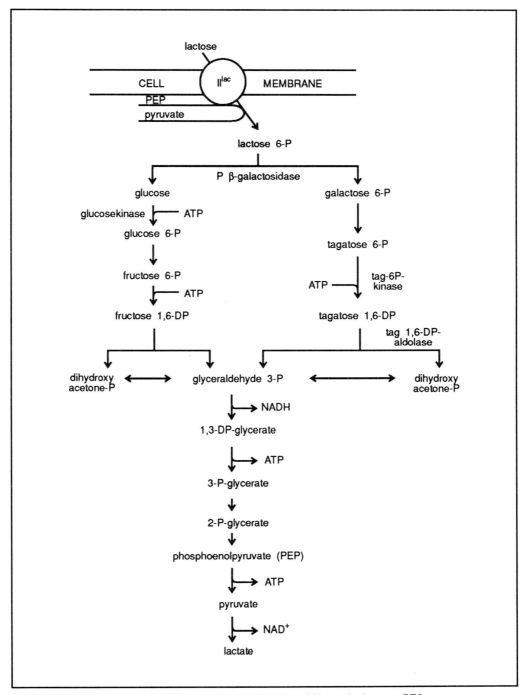

Figure 3.2a Transportation and utilisation of lactose in lactic acid bacteria; Lactose PTS system.

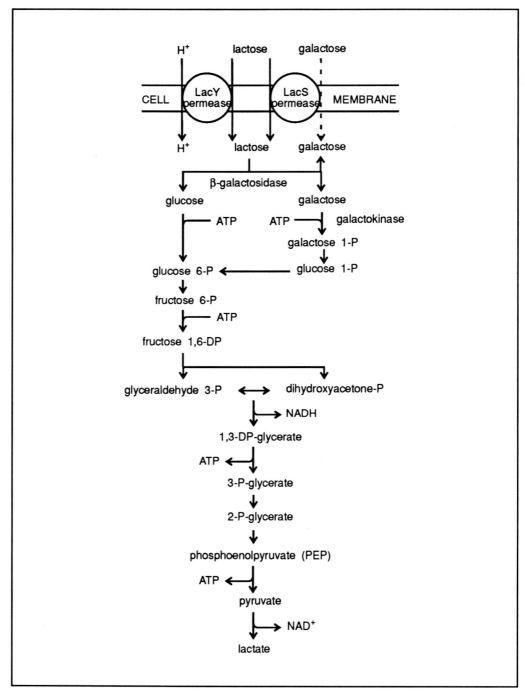

Figure 3.2b Transportation and utilisation of lactose in lactic acid bacteria; system using two different lactose permeases (Lac Y and Lac S).

In most other lactic acid bacteria such as *Lb acidophilus*, lactose is not modified during transport into the cells, instead, uptake is believed to be mediated by a lactose permease. Once within the cell the disaccharide is hydrolysed by a β-galactosidase. The

PS

biochemistry and genetic control of this so-called lacY permease system (PS) is well known from historical studies in the model bacterium *Escherichia coli*.

Recent studies have indicated that in the yoghurt bacteria *Lb. bulgaricus* and *St. thermophilus* the transport of lactose is mediated by a different permease, the LacS permease, and is accompanied by the excretion of galactose (this is not metabolised by these bacteria because they have low activities of the enzyme galactokinase).

In both *Leuconostoc* and *Lb. helveticus* , the galactose moiety (part) of lactose is metabolised. Both lactic acid bacteria contain β-galactosidase and it is conceivable that lactose is transported via a LacS lactose permease.

SAQ 3.1

In column A lactose transport systems are listed. Match each with the appropriate lactic acid bacterium in column B, and the appropriate characteristic in column C.

A	B	C
lacY permease	*Lb. acidophilus*	galactose utilised
PTS	*L. lactis*	galactose excreted
lacS permease	*Lb. bulgaricus*	PEP-dependent

Lactococcus spp., *Lb. bulgaricus* and *St. thermophilus* show homolactic fermentation resulting exclusive in the production of lactate. The stereospecificity of the lactate produced varies since all lactococci and *St. thermophilus* produce L-lactate and *Lb. bulgaricus* forms the D-isomer of lactate. Similarly, lactose is fermented by *Lb. helveticus* in a homolactic fashion presumably via the pathway shown in Figure 3.2. In contrast, in *Leuconostoc spp.* hexose fermentation is heterofermentative and occurs via the phosphoketolase pathway in which 1 mol each of lactate, ethanol and CO_2 are produced per mol substrate metabolised. Finally, *Leuconostoc* species produce D-lactate, whereas by *Lb. helveticus* a racemic mixture of both D- and L- isomers of lactate are formed.

For the production of cheeses with an open structure such as Emmenthaler, use is made of *Propionibacterium spp.* such as *Pr. shermanii* and *Pr. freudenreichii*. The organisms convert lactate to propionate and other products via the following reaction:

$$3 \text{ lactate} \rightarrow 2 \text{ propionate} + \text{acetate} + CO_2 + H_2O$$

The ratio of propionate to acetate varies with different strains. The CO_2 (gas) produced is responsible for the production of the open structure (holes or eyes) in the cheese.

Citrate metabolism

diacetyl

The main citrate-utilising bacteria are *L. lactis* subsp. *lactis* var. *diacetylactis* and *Leuconostoc spp.* such as *Lc. lactis*, *Lc. cremoris* and *Lc. plaetis*. *L.lactis* subsp *lactis* var *diacetylactis* looks a little complex for a name for an organism. Let us look at it a little more closely. L stands for the genus *Lactococcus*, *lactis* is the species name. This species however is divided into a number of subspecies. One of these is also called *lactis*, hence the second *lactis* in the name. There are several varieties of this subspecies. One of these is *diacetylactis*. The metabolism of citrate by these mesophilic lactic acid bacteria is shown in Figure 3.3. The important intermediate in this metabolism is α-acetolactate, which is decarboxylated under various conditions into diacetyl, the important flavour compound in butter, buttermilk, cottage cheese and quark. In addition, the CO_2 produced contributes to the formation of eyes in cheese such as Gouda.

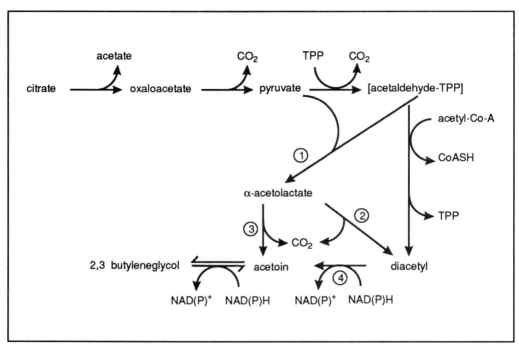

Figure 3.3 Citrate metabolism in lactic acid bacteria. TPP = thiamine pyrophosphate, CoASH = Coenzyme A. 1) acetolactate synthase, 2) acetolactate decarboxylase, 3) chemical oxidation, 4) diacyl reductase.

∏   What type of organism do you think could be used to give a higher yield of diacetyl? (Look at Figure 3.3 and think about the role of diacetyl in this metabolism).

Diacetyl is an intermediate product and is converted to acetoin. A mutant organism with a defective enzyme for conversion of diacetyl to acetoin would accumulate diacetyl (it would be unable to convert it to acetoin).

Protein metabolism

fastidious   The starter bacteria used in milk fermentation are nutritionally very fastidious (that is they have complex requirements) and numerous amino acids are either stimulatory or essential for their growth. Although milk is rich in protein, it does not contain sufficient free amino acids to support rapid growth, so some starter organisms produce
proteolytic   proteolytic enzymes that generate free amino acids from milk protein. The proteolytic
enzymes   capacity of starter cultures varies greatly between strains and species of bacteria used. In many cases therefore, mixed cultures consisting of proteolytic and non-proteolytic strains are used, the latter being dependent on the presence of the proteolytic strain(s) to provide the necessary amino acids.

An important reason for the use of micro-organisms other than lactic acid bacteria is to increase the proteolytic degradation of the milk protein and hence stimulate diverse flavour development. Flavour may originate from the individual amino acids themselves but may also develop via decarboxylation, de- and trans-amination and desulphurylation of their derivatives. These activities generate a variety of amines, aldehydes, alcohols and various sulphur compounds. An example is the use of *Propionibacterium spp.* The organisms are not just associated with propionate fermentation, but also contribute to proteolysis by the production of proline-containing peptides. This amino acid is considered to be an important contributor to the sweetness of Swiss-type cheeses. Another example is ammonia, which is an important factor in the flavour of Camembert cheese. Moulds such as *Penicillium spp.* contribute to the overall proteolysis but are also reported to produce hydrogen sulphide, dimethyl sulphide and methane thiol from methionine. Finally, the main flavour of yoghurt is acetaldehyde that is mainly derived from the degradation of threonine via the action of the threonine aldolase of *Lb. bulgaricus.*

∏ Can you name an amino acid (other than proline) which contributes to the flavour of foods? You might find it helpful to think of the foods you find on the supermarket shelves. Some have quite considerable amount of an amino acid added, not primarily to enhance the nutritional value of the food but to improve its flavour.

flavour enhancer

Glutamic acid is widely used in the cooking of oriental foods. It is manufactured by microbial fermentation, and produced as the sodium salt, to give monosodium glutamate (MSG). This is called a flavour enhancer. We will discuss its manufacture in Chapter 8.

Lipolysis

Lactic acid bacteria used as starter cultures are only weekly lipolytic. This is particularly true of the mesophilic starters used in the production of cultured butter (see Section 3.3.3). In contrast, micro-organisms constituting the secondary flora contribute more significantly to the hydrolysis of milk fat and the production of free fatty acids.

3.3.3 Cultured butter and buttermilk

Cultured butter is very popular in various countries, for example, The Netherlands. It is traditionally made from milk fat to which a mesophilic starter culture has been added which is capable of generating the flavour compound diacetyl by metabolising citrate. Milk is centrifuged, by which the fat (cream) is separated from the low-fat milk. The cream is pasteurised and inoculated with starter culture. After incubation the 'ripened

cream butter

sour buttermilk

cream' is added to a churn and mixed. As a result of the high acidity, the cream separates into cream butter and sour buttermilk. The sour buttermilk is the by-product and has limited use as it has high acidity. To overcome this problem a process has been developed in The Netherlands to produce butter without the formation of any sour buttermilk. In this process lactose-reduced whey is pasteurised, inoculated with *Lb. helveticus* and fermented to produce lactic acid. The culture is centrifuged to remove the cells and then lactic acid is separated and recovered by ultrafiltration. At the same time, low-fat milk is pasteurised and inoculated with a starter culture designed to produce

sour aromatic butter

high levels of aroma compounds (acetolactate and diacetyl) and lactic acid. The pasteurised cream is churned first, resulting in the separation of a valuable sweet (because of the residual lactose) by-product ('sweet buttermilk') that can be used for other purposes. Subsequently, the lactic acid and 'aromatic starter' are added to the cream which is further churned. The resulting butter is known as sour aromatic butter that cannot be distinguished from the traditional ripened cream butter.

flow diagram

These processes are described in Figure 3.4. This type of diagram is called a flow diagram because the process is represented from start to finish with the order in which processes are carried out shown by arrows. We are sure you will be familiar with this idea. If you are not, compare the description given with the diagram, to see how it operates. The steps in the process (sometimes called unit processes) are represented by a drawing of the piece of equipment used, or simply by a box. Materials put in or taken out (inputs and outputs) are not in boxes. We will be using this format throughout this text. However, elsewhere you might find flow diagrams represented differently.

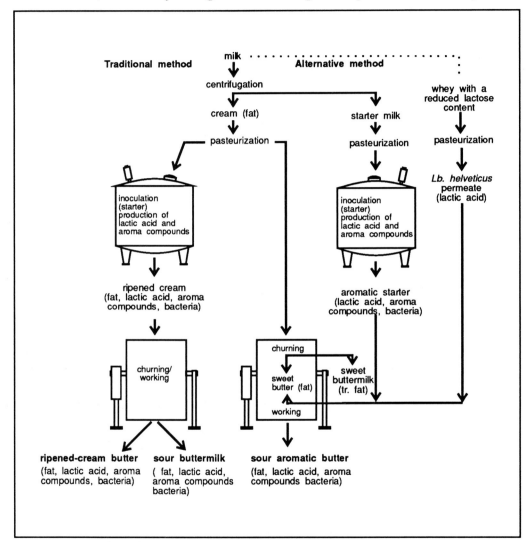

Figure 3.4 The production of cultured butter and buttermilk.

As this process eliminated the supply of buttermilk, an alternative buttermilk production method was developed by fermenting milk with a mesophilic starter culture of *L. lactis* and *Leuconostoc spp.* (see Table 3.1). The main flavour compound in this product is diacetyl produced by the citrate metabolism of the starter cultures used.

3.3.4 Yoghurt and other fermented milks

Yoghurt is a fermented milk product that is produced at temperatures between 30 - 50°C by inoculating milk with a stable mixed culture of two thermophilic lactic acid bacteria, *St. thermophilus* and *Lb. bulgaricus*. The characteristic yoghurt flavour, caused by the aroma compound acetaldehyde, is produced by *Lb. bulgaricus* as part of its protein metabolism as described earlier. *St. thermophilus* is involved in the generation of the fresh, acid taste of the produce. Both yoghurt bacteria may produce extracellular polymers that contribute to the characteristic viscosity of the product. The stability of the mixed strain culture results from the fact that interactions favourable to both lactic acid bacteria occur. These interactions, also called protocooperation, are shown in Figure 3.5 and include:

- the proteolytic degradation of the milk protein by *Lb. bulgaricus*;

- the production of small amounts of formic acid that are formed at low oxygen concentrations by *St. thermophilus*.

acetaldehyde

proto-cooperation

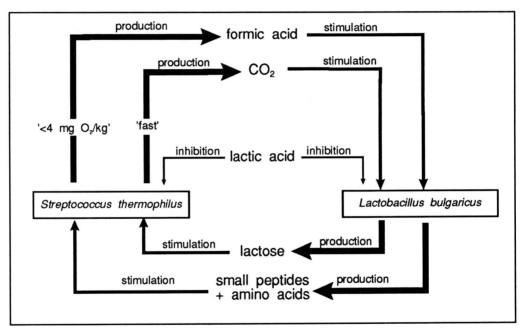

Figure 3.5 Protocooperation by yoghurt bacteria.

In the production of yoghurt, the milk is first of all standardised to a fat content of 0.5-3% and is fortified with 14-16% milk solids (dried milk). It is warmed to 50-60°C and homogenised to disperse the fat. The milk is pasteurised by heating to 85°C for 30 minutes, 95°C for 5-10 minutes or 120°C for 5 seconds. After pasteurisation the milk is cooled to incubation temperature and inoculated with starter cultures. Incubation is at 30°C or 45°C, the incubation period being shorter at the higher temperature. Set yoghurt is packed into its retail container after inoculation, and allowed to ferment within its container. The containers are then stored in a refrigerated state. Stirred yoghurt is incubated in bulk after inoculation, homogenised and then packed into containers. Fruit is added at the time of packing in the retail container for the production of fruit yoghurt.

set yoghurt

stirred yoghurt

SAQ 3.2	Presented below is a flow diagram representing the production of plain yoghurts. From the list provided, fit in inputs and unit processes at points marked * and complete the diagram.

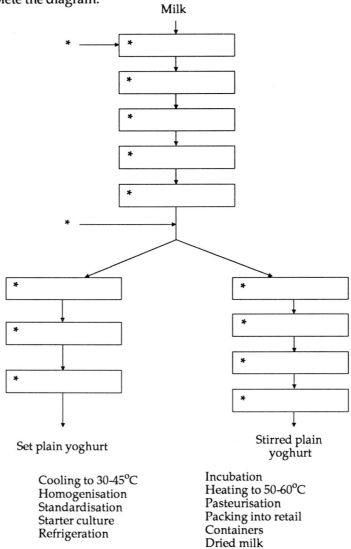

Set plain yoghurt

Stirred plain yoghurt

Cooling to 30-45°C
Homogenisation
Standardisation
Starter culture
Refrigeration

Incubation
Heating to 50-60°C
Pasteurisation
Packing into retail
Containers
Dried milk

Drinking yoghurt is categorised as a stirred yoghurt of low viscosity and this product is consumed as a refreshing drink. Frozen yoghurt, which is based on natural yoghurt, is gaining increasing popularity as an alternative for ice cream.

Continuous production of yoghurt has been developed and patented but the process has, so far, not been used for industrial production of yoghurt. In this process standardised pasteurised milk is put into a mixed vessel, held at 45°C, and inoculated with starter culture. When the pH drops to 5.7, the process is operated continuously at a set volume and with the pH maintained at 5.7 by the in-flow of fresh milk. The out-flowing fermented milk (at pH 5.7) is cooled to 37°C, allowed to coagulate, then treated in the usual way.

∏ Would the continuous yoghurt process produce set yoghurt or stirred yoghurt? (Think back to the distinction we made between set and stirred yoghurt).

The process uses continuous fermentation prior to packaging. The yoghurt is fermented prior to packaging and thus this process produces stirred yoghurt.

One of the main problems associated with yoghurt production is continued acidification during storage. This can be significant, even at refrigeration temperatures. This problem can be overcome by pasteurising the yoghurt after the fermentation has taken place. However many consumers prefer unpasteurised or 'live' yoghurt.

∏ In which ways do you think that genetic manipulation of yoghurt bacteria could be used to overcome the problem of acid production during refrigerated storage? (There are several alternatives - think about the differences between the temperatures of incubation and storage of yoghurt).

Strains (or mutants) having reduced capacity for acid production at low temperatures (compared to production strains) could be isolated. The genes for lactose permeases and β-galactosidase have been isolated and cloned. It may be possible to alter these (by site-directed mutagenesis) so that they become cold-labile (inoperative at cold temperatures). An organism possessing such cold-labile enzymes could not produce acid at refrigeration temperatures. At present, cloning vectors necessary for such genetic engineering procedures are well developed only for *St. thermophilus*.

health and new yoghurts

Recently a variety of (fermented) milk products have been developed that contain *Lb. acidophilus* and *Bifidobacterium spp.* (mainly *B. bifidus*) as sole cultures or in combination with traditional yoghurt or buttermilk cultures. Although some of these thermophilic strains may colonise the human intestinal tract, it remains to be determined how efficiently the yoghurt bacteria compete with the endogenous intestinal microflora. This is quite an important issue because intestinal function will, at least in part, be influenced by the microflora present. Since we do not know how effective yoghurt micro-organisms are at colonising the human intestinal tract, the claims regarding improved health due to improved intestinal function associated with the consumption of these products needs more thorough investigation. An indisputable fact is that these products and other fermented milk products such as standard yoghurt and buttermilk do contain reduced contents of lactose, and are therefore suitable for consumption by lactose-intolerant individuals.

3.3.5 Cheese

dehydration

Cheese manufacture is essentially a dehydration process in which the milk protein (casein) and fat are concentrated between six and twelve fold, (depending on the variety). Although the manufacturing protocols for individual varieties may differ, the basic steps common to most varieties are:

- acidification of the milk, which is a consequence of the lactose fermentation by lactic acid bacteria;

- coagulation of the casein by the concerted action of limited proteolysis and acidification. The proteolysis is initiated by added animal or fungal proteinases (rennets). The coagulated caseins form a gel which entraps any fat present;

SAQ 3.3	We have described the production of yoghurt as being dependent upon a mixed culture of two thermophilic lactic acid bacteria. It would be simpler, if a single culture could be used in order for us to remove the dependence on *Lb. bulgaricus*. Which one of the following genes should be cloned and expressed in *St. thermophilus*, in order for it to be independent (see Figure 3.5).

1) β-galactosidase.

2) LacY permease.

3) Cold-labile β-galactosidase.

4) Lipase.

5) Protease.

• dehydration, which involves the post-coagulation treatments that cut or break the gel, resulting in expulsion of the whey (which contains most of the water, milk sugar and milk proteins other than casein;

• shaping, which contributes to the final appearance of the product;

• salting, which has various effects including the reduction of microbial growth and activity (by lowering the A_w), the control of the activity of enzymes in cheese and the formation of physical changes in cheese proteins. These effects result in a contribution of salt to the preservation of the cheese and its flavour development;

• ripening, in which an extremely complex series of biochemical changes occur through the catalytic action of the coagulant, starter bacteria and their enzymes, secondary microflora and their enzymes and endogenous milk enzymes. These changes include glycolysis, proteolysis and lipolysis.

∏ Draw yourself a flow diagram of the overall stages in cheese production

soft and hard cheeses

The number of cheese varieties is enormous, and various parameters have been used to discriminate between them. One such is the amount of water contained in the final product. This criterion has been used in classifying the cheeses presented in Table 3.1 into soft (55% water), semi-soft (43-55% water) and hard (water) cheeses. We will discuss the production of two of the most important cheeses, Cheddar and Gouda in more detail in Chapters 4 and 5. For Cheddar type cheeses, various strains of *Lactococcus lactis* are normally used as starters. With other cheeses, other organisms may be used (Table 3.1). These provide the cheeses with their characteristic flavour and texture.

3.4 Acid-fermented vegetable products

3.4.1 Sauerkraut

importance of salt

The major fermented vegetable foods in Europe and the USA are cabbage (sauerkraut) and pickles (for example cucumbers and olives). The principles of all processes are more or less the same. A flow chart of sauerkraut production is shown in Figure 3.6. After harvesting, the cabbage is shredded and pre-salted before storing in a tank under anaerobic conditions. Salt is used for several reasons. It reduces the water activity (A_w) thereby creating selective conditions in favour of endogenous lactic acid bacteria. In addition, it promotes leakage of sugars from plant cells and consequently the liberation of fermentable sugars necessary for growth and fermentation by lactic acid bacteria. The cabbage is packed in vats or bins, so promoting anaerobiosis.

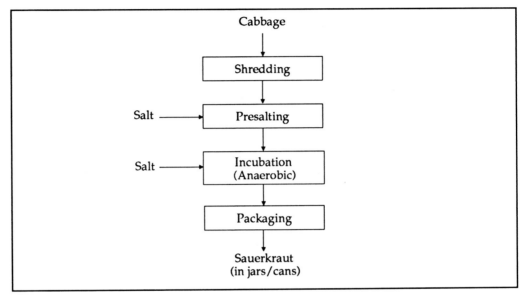

Figure 3.6 The production of sauerkraut.

The sequence of microbial activity is very similar in all acid-based vegetable fermentations (Figure 3.7). Gram-negative bacteria are the most common organisms found on vegetables (ca 10^4-10^6/g plant tissue) due to their less-complex nutritional demands. Lactic acid bacteria are generally present in low numbers (10-10^2/g). Stimulation of the growth of lactic acid bacteria (by anaerobiosis and low A_w) leads to a lowering of pH.

Gram-negative followed by lactic acid fermenters

We can divide the effect on microbial growth into a number of phases. Phase 1 is characterised by a repression of the activity of the Gram-negative microflora. This repression depends on the rate of lactic acid formation and pH decrease. These will depend upon the level of fermentable sugars and the buffering capacity (respectively), as well as on the type of lactic acid bacteria present (homo- or heterofermentative). The type of fermentation will also determine the flavour of the product. In cabbage fermentation, four types of lactic acid bacteria dominate this phase, namely *Leuconostoc mesenteroides* (heterofermentative), *Lactobacillus brevis* (heterofermentative) and the homofermentative *Lactobacillus plantarum* and *Pediococcus cerevisiae*

effects of temperature and salt on lactic acid fermenters

Temperature and salt concentration control the type of lactic acid bacteria predominating in cabbage fermentation in the primary stage (Figure 3.7) and secondary stage of fermentation (see Figure 3.8). At a salt concentration of 2.25% and at relatively low temperatures (7.5°C), both phases tend to be dominated by heterofermentative *Leuconostoc mesenteroides* (a lactic acid bacterium with a very short lag and generation time). Under these conditions, sauerkraut superior in flavour and colour is produced. At higher incubation temperatures the more acid-tolerant *Lactobacillus* and *Pediococcus* species dominate in the primary and secondary phase respectively.

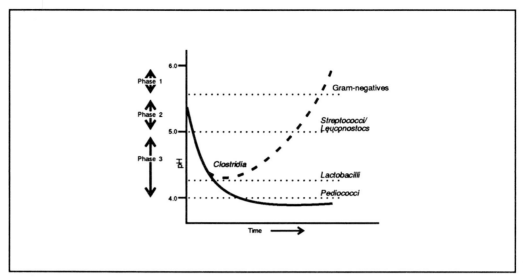

Figure 3.7 Phases in the fermentation of vegetables eg sauerkraut.

Higher incubation temperatures are favourable for *Lactobacillus* and *Pediococcus* species, and result ultimately in a fully homofermentative process (acid sauerkraut). Temperature is therefore the main controlling factor for quality of sauerkraut due to the different metabolisms of the bacteria involved. The effect of temperature on the development of organisms are shown in Figure 3.8.

Π Which of the temperatures used to produce the data in Figure 3.8 showed the lowest acid formation?

The lowest percentage of acid is formed at 7.5°C.

For a stable product, pH values of 4.0-4.5 are required (the lower the A_w, the higher the pH can be). For a fully stable product all the sugars in the product should be fermented.

3.4.2 Cucumbers and olives

Cucumber and olive fermentations are performed at far higher salt concentrations compared to sauerkraut and therefore an altered lactic acid bacterial activity occurs. The sequence of microbial phases is however very similar to sauerkraut production. A consequence of salt concentrations of 5-8% (as used in olive fermentations) is that *Leuconostoc mesenteroides* never predominates in the primary fermentation phase and at 8% salt concentration may even be absent.

At this salt level *Lactobacillus plantarum* generally predominates with *Pediococcus acidilactici*, *Pediococcus pentosaceus* and *Lactobacillus brevis* present in lower numbers.

Problems and new developments

Spoilage of sauerkraut is generally avoided by maintaining anaerobic conditions and low pH. This eliminates potential problems that could otherwise be caused by aerobic microbes.

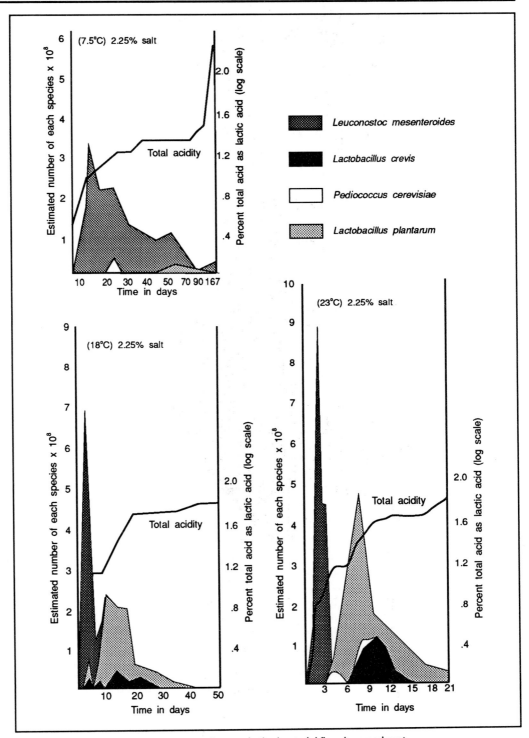

Figure 3.8 The effect of temperature on changes in the bacterial flora in sauerkraut.

∏   What spoilage problems do you think that aerobic organisms could cause in
     sauerkraut? (What can live heterotrophically on plant materials at low pH?).

bloater
damage

Filamentous fungi could grow in the low pH found in sauerkraut if oxygen is present.
This can result in: off 'mouldy' flavour; growth of visible (coloured) fungal biomass on
the product; loss of acidity due to the formation of alkaline products (such as ammonia);
softening of the product due to pectinase production (see Chapter 7).

gassy, floater
or fish-eye
spoilage

A common problem of whole fruits such as cucumber and olives is gaseous spoilage.
This causes so called 'bloater' damage and 'gassy', 'floater' or 'fish-eye' spoilage. Many
gas forming micro-organisms may be involved in this process.

∏   Could the lactic acid bacteria contribute to gaseous spoilage? (Look back at Figure
     3.1 for a clue).

Yes, heterolactic fermentation results in the formation of CO_2 gas. Organisms such as
Leuconostoc mesenteroides could contribute to this type of spoilage if allowed to grow
uncontrollably. Other gas-forming organisms which could cause such spoilage are
anaerobic bacteria and yeasts.

Controlled fermentation processes, ensuring the development of a desirable lactic acid
bacterial flora by use of started cultures is therefore advisable in whole fruit
fermentations.

New biotechnological approaches in this area are directed to the improvement of starter
cultures. The genetics of lactic acid bacteria has been well studied in the dairy industry,
and we will consider it in detail in Chapter 4. At present, the development of starters is
largely confined to the production of pickles, as the complex interrelated reactions
occurring during sauerkraut fermentation are difficult to reproduce. Essentially what is
required is more research.

3.5 Acid-fermented meat products

The slaughtering and butchering process contaminates the initially sterile tissue (meat)
with many Gram-negative micro-organisms such as *E. coli*, *Salmonella spp.* and
Pseudomonas spp. and Gram-positive micro-organisms. This flora is associated with
intestines or the environment. Relatively few lactic acid bacteria are initially present. As
in vegetable fermentation, a micro-environment has to be created to control the
outgrowth of the Gram-negative spoilage flora, as well as some pathogens such as
Clostridium perfringens. As with sauerkraut, control is by creating anaerobic conditions
and lowering the A_w by salting, curing (adding salt and compounds such as nitrite)
and/or drying. Often starter cultures are also used (see Table 3.2).

We would not expect you to remember all of these names - but it would be helpful to
remember the main ones. To help you, do the following intext activity.

∏   Using Table 3.2 work out which species is mentioned the most. Is the use of this
     organism confined to the one type of meat? Which other species are described for
     more than one type of meat?

Product	Bacteria
Semi-dry fermented meat sausages:	
Lebanon bologna	Mixture of *Pediococcus cerevisiae* and *L. plantarum*
Summer sausages	*P. cerevisiae* or *P. cerevisiae/L. plantarum*
Cerevelat	*P. cerevisiae* or *P. cerevisiae/L. plantarum*
Thuringer	*P. cerevisiae*
Teewurst	*Lactobacillus species*
Pork roll	*P. cerevisiae*
Dry fermented meat sausages:	
Pepperoni	*P. cerevisiae/L. plantarum*
Dry sausage	*P. cerevisiae*
European dry sausage	*Micrococcus species* or *Micrococcus/Lactobacillus species*
Salami	*Micrococcus/Lactobacillus species* or *L. plantarum*
Hard salami, Genoa	*Micrococcus species, Micrococcus species/ P. cerevisiae; Micrococcus species/L. plantarum*
Fermented poultry sausages:	
Semi dry turkey sausage	*P. cerevisiae*
Dry turkey sausage	*P. cerevisiae* or *P. cerevisiae/L. plantarum*
Pickles:	
Carrots	*L. plantarum/L. brevis/ Leuconostoc mesenteroides/ Pediococcus cerevisiae*
Cucumbers	*P. cerevisiae/L. plantarum/ L. brevis; P. cerevisiae/ L. plantarum; L. plantarum*
Mixed vegtables, green tomatoes, hot cherry peppers	*P. cerevisiae/L. plantarum; L. plantarum*
Olives	*L. plantarum*
Sauerkraut	*L. plantarum*

Table 3.2 Bacteria used as starter cultures in meat, poultry and vegetable products.
NB species names separated by / indicate a mixture of organisms used.

The organisms mentioned the most is *Pediococcus cervisiae*. It is used in many different fermented meat products. *Lactobacillus plantarum* is also used in the fermentation of different meats.

back-slopping
technique

In some instances the starters used are similar to those used in fermented vegetable products. Choice of starter is partly limited by the fermentation rate required. In traditional processes, rather than use cultured starters, the 'back-slopping' technique is used, in which some of the fermented sausage is used as an inoculum for the next fermentation. However this technique is not recommended as Good Manufacturing Practice (GMP). Let us see if we can work out why this is so.

∏ Why is the growth of *Clostridium perfringens* in food undesirable? (Look back at the last two paragraphs or so).

The organism can cause food poisoning if allowed to grow in food and the food is eaten.

∏ Why do you think 'back slopping' is not Good Manufacturing Practice?

If food spoilage or food poisoning micro-organisms develop during one fermentation (due to inadequate reduction in A_w or pH or oxygen levels), then these organisms will be inoculated into the next batch of processed meat. They may be able to outgrow the required (starter) organisms and reach high numbers before the pH is reduced by lactic acid production during successive meat fermentations. Such a product may have abnormal organoleptic properties or pose a risk of food poisoning if eaten.

A variety of fermented meat products exists (Table 3.2). The majority can be classified as either dry or semi-dry fermented sausage. These have a respective final moisture content of 25-45% (A_w: ca 0.91) and 45-50% (A_w: ca 0.95). For production of these sausages, meat curing compounds and usually a starter culture are mixed at low temperatures (minus 7°C - minus 1°C). After stuffing in casings and incubation (fermentation) at 21-43°C, drying follows at 10-21°C depending on the type of product.

The fermentation process is not only important in developing the stability of the end product but also the texture, colour and flavour. Acid formation also makes the final drying process easier. Starter cultures therefore have to be selected carefully, in order to produce the required activities.

3.5.1 Dry sausage

A clear distinction exists between meat fermentation technology for dry sausage production in the USA and many European countries.

fast acid
fermentation

In the USA, and in some processes in European countries (such as The Netherlands), fermentation processes are directed at producing a fast acid fermentation to inhibit Gram-negative spoilage organisms and *Staphylococcus aureus* strains. Nitrite is added together with salt in the 'curing' mixture, and is necessary to produce the colour of the product (which it does by chemical reaction with myoglobin in the meat). In the USA, for fermentation at 40°C *Pediococcus acidilactici* is used as starter. At lower temperatures *Pediococcus pentosaceus* is used. In The Netherlands *Lb. plantarum* is used for fast fermentations, also develop predominantly in natural meat fermentations. *Lactobacillus plantarum* has been judged as starter with the best feremntation characteristics (homofermentative, tolerates more than 9% salt and high acidic conditions). The main part of flavour formation has been contributed by lactic acid in these fast fermentation processes. The product has a final pH of 4.6-5.1.

Slower fermentations at 24°C are carried out in many European countries. In these processes, nitrate is added with salt as a curing agent (instead of nitrite). After the lactic acid fermentation the chopped meat is packed into metal pans to a depth of about 20cm and incubated at refrigeration temperatures for a period of 2-7 days. Various species of *Micrococcus* and *Staphylococcus* (already present in the meat) covert the nitrate to nitrite. In the future such organisms may prove useful as starter cultures. This type of process is called pancuring, because of the use of the pans to hold the meat.

pancuring

New developments

As is the case of sauerkraut fermentation at low temperatures, the endogenous microflora and starter organisms in the pancuring process have a wider role than lactic acid production alone. Lipolytic and proteolytic enzymes produce important flavour compounds which probably produce characteristic organoleptic properties in dry sausage. In fast fermentations, rapid acid production alone is the main goal. Strain improvement in starter cultures is the important area of development. Prevention of too much acid production and increasing the lipolytic and/or proteolytic activity are the main aims.

∏ Do you think that it would be easier to apply genetic engineering to the fast fermentation or pancuring system of production? (Think about the roles of the starter culture in the two types of process).

The fast fermentation process relies on lactic acid fermentation alone. This would be easier to manipulate on its own than lactic acid production and nitrate reduction in combination (as would be necessary in the pancuring process).

| **SAQ 3.4** | Which of the following completions to the statement is/are true? |

Lactic acid alone cannot bring about preservation of sauerkraut because:

1) filamentous fungi could still grow and cause spoilage;

2) anaerobic bacteria could still grow and cause spoilage;

3) yeasts could still grow and cause spoilage.

SAQ 3.5	Which of the following completions to the statement is/are true?

Lactic acid contributes to the preservation of acid-fermented foods by:

1) helping to prevent the growth of spoilage organisms which could otherwise produce enzymes which degrade the food;

2) helping to prevent the growth of spoilage organisms which could otherwise produce metabolites which spoil the food;

3) helping to prevent the growth of lactic acid bacteria which could otherwise produce enzymes which degrade the food;

4) helping to prevent the activity of endogenous enzymes which would otherwise degrade the food. |

SAQ 3.6	In column A are listed acid-fermented products. Match each product with the organism in column B with which it is associated, and with the conditions in column C (neither, either or both) which also contribute to preservation.

A	B	C
yoghurt	*Streptococcus thermophilus*	low water activity/
sauerkraut	*Pedioccus acidilactici*	low water content
cheddar cheese	*Leuconostoc mesenteroides*	
dry sausage	*Lactococcus lactis*	

3.6 Fermented Oriental foods

The development of fermented foods in the Orient has differed in many ways from the acid-based fermentations in Western countries. Although somewhat similar with respect to preservation, the improvement of nutritional value and (especially) the flavour are more important characteristics in oriental food fermentation processes. For instance alternatives have been developed for meat and meaty flavours by fermentation of vegetable products such as soya. Such developments have been stimulated by the Buddist religion which forbids meat consumption.

solid-substrates Many oriental food fermentations are based on a solid-state (or solid-substrate) fermentation technology. Solids are fermented by micro-organisms without the presence of free water (in contrast to submerged fermentations, generally practised in Europe). Water content varies however from 20-80% depending on the water binding properties of the substrate. The predominant use of fungi in these processes is a logical consequence of this condition. Several types of solid-state fermented food can be categorised:

A   food initially fermented by fungi and subsequently followed by a brine process (as in soya sauce production);

B   meat flavour pastes produced in a one-step bacterial fermentation;

C   fermented foods produced by an acid soaking process and subsequently a fungal fermentation (as in tempeh production);

D   doughs fermented by lactic acid bacteria;

E   alcoholic foods produced by fermentation with yeasts or other fungi.

Some examples of fermented food in these different categories are shown in Table 3.3. This list is however far from complete, due to the enormous numbers of Oriental and African products.

CATEGORIES				
A	B	C	D	E
shoyu 1	thua nao 1	tempeh bongkrek 4	idli 2	tape ketella 6
hamma-natto 1	natto 1	oncom 5	puto 2	tape ketan 2
tauco 1		tempeh kedale 1	minchin 3	lao chao 2
doenjang 1		tempeh gembus 1	jalebies 3	
miso 1,2		tempeh koro 7		
ketjap 1				

Table 3.3 Some examples of the different categories of solid state fermented Oriental foods. Based on: 1) Soya, 2) Rice, 3) Wheat, 4) Coconut, 5) Peanut, 6) Cassava, 7) Beans. See text for categories (A-E).

∏   Why do you think fungi are used most often in solid-substrate fermentations?

unavailable substrates

The solid-substrate fermentations are based on solid material such as wheat, rice and soya beans. The conditions are aerobic, with low A_w (due to the low moisture content). Furthermore the nutrients in the substrate are in complex forms (such as starch and proteins). Unless organisms produce enzymes capable of degrading these substrate (such as amylases and proteases) then the substrate cannot be used (in which case it is said to be unavailable). Fungi are generally capable of growth in such conditions, as they produce a variety of degradative enzymes.

In production of soya sauce (shoyu), miso and to some extent also for hamma natto, tauco, doenjang, sufu and kochujang, similar steps in production technology can be observed. This technology can be divided in three phases: 1) koji production; 2) a brine fermentation and 3) a refining phase.

The koji production (phase 1) is a unique technology using solid-state fermentation for production of enzyme-rich substrate, necessary as the starting material for manufacturing many Oriental foods, also wines (sake), spirits (shochu) and vinegar (yonezu).

3.6.1 The koji process

The koji process is akin to the malting process used for beer and spirit production in Western countries. In brewing, the substrates present in malt (such as starch and protein) are unavailable to yeasts (yeasts cannot use them as substrate). By germinating the barley during malting, endogenous amylases and proteases are produced. During *mashing* mashing (soaking the malt) these enzymes degrade starch and proteins to maltose and amino acids. These nutrients can be used by yeast (ie they are available).

In koji production, the enzymes necessary to convert the substrate to forms available to yeasts or bacteria are produced by a fungus. The substrate (soya beans, wheat or rice - or mixtures of these) is first of all cooked by steaming. This causes hydration (adds water to the substrate), helps to render the starch into a soluble form and reduces the number of contaminating organisms (which could otherwise outgrow the starter culture). After cooling, the mixture is inoculated with the spores of *Aspergillus oryzae* or *Aspergillus sojae*. The inoculated mass is placed in shallow pans or baskets and is incubated in a humid environment (to prevent drying) at about 22°C for 3-4 days.

During the fermentation period the fungus produces amylases and proteases, as well as pectinases which cause maceration (separation of the cells - see Chapter 6). The solid-substrate fermentation process leads to balanced production of these enzymes (that is in quantities suitable for the required purpose). In submerged fermentation, the fungi would produce different quantities of these enzymes, which would lead to products with different organoleptic properties. Some disadvantages should be mentioned regarding solid state fermentation in general. Temperature and humidity control are very important parameters in koji production as in all other solid-state fermentation processes. Temperatures influence the balance of enzymes produced and even inactivation occurs at inappropriate control. Compared to submerged fermentation processes the process control and regulation is relatively under-developed.

Recent biotechnological improvements have been made in mainly process technology of koji production. Continuous cookers, automated inoculators, mixers, large perforated vats in closed chambers equipped with forced air devices and temperature controls are nowadays used frequently (Figure 3.9).

3.6.2 Soya sauce, miso and sufu production

In soya sauce production, koji is made from wheat and soyabeans which are finely ground (meal), rolled (flakes), or crushed (grits) and mixed with brine to form a moromi mash. The salt concentration is adjusted to 16-19%. Such a high salt concentration releases the mycelium-bound enzymes and inactivates the koji starter and many contaminating micro-organisms. The enzymes of the koji degrade the carbohydrate and proteins of the raw materials to sugars, peptides and amino acids (forming the nutrients required for a second fermentation). The mash is filtered (strained) and placed in vats. As in acid food fermentations, a sequence of microbial activities occurs during mash fermentation due to lactic acid bacterial activity. *Pediococcus halophilus* dominates the first phase. The final low pH-values lead to favoured conditions for outgrowth of the osmophilic yeast *Saccharomyces rouxii* and a vigorous alcoholic fermentation follows. *Torulopsis* species (a yeast) dominate the last phase of brine fermentation due to a better resistance to high dissolved nitrogen concentrations and alkylphenols and aromatic alcohols formed by yeast activity.

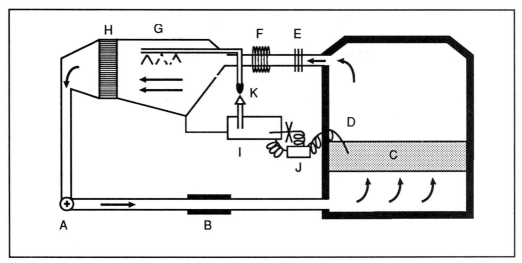

Figure 3.9 A solid-substrate bioreactor used in the koji process. A - fan, B - heater; C - fermenting solid mass, D - thermometer, E - air filter, F - damper, G - shower, H - drain eliminator, I - water bath, J temperature regulator, K - pump.

In the lengthy secondary fermentation phase which follows, a mixed population of micro-organisms form aromatic substances such as acids, alcohols, esters and several phenolic compounds (4-ethylguaiacol, 4-ethylphenol and 2-phenylethanol). In addition, the glutaminases produced during koji production contributes to flavour development by forming glutamate from amino acids. More than 100 aroma compounds have been identified in soya sauce. As can be expected the flavour of soya sauce can be regulated by adjusting wheat-soya bean ratios.

SAQ 3.7	Draw a flow diagram to show the production of soya sauce from wheat and soya beans.

Miso production differs in two essential ways from soya sauce production. Firstly, in miso production rice alone is fermented by *Aspergillus spp.* to produce koji. Whole soya beans are subsequently added to the rice koji. As a consequence, the degree of hydrolysis of soya proteins is lower. Secondly, the second fermentation in miso is an anaerobic fermentation of a paste (rather than a liquid brine fermentation). The microbial activity is however basically the same as in soya sauce fermentation.

New biotechnological advancements in soya sauce and miso production have been made recently by improvements to the technology of solid-substrate fermentation in the production of koji. The complex nature of the flavour composition and the mixed culture character of the second fermentation complicate use of pure starter cultures. For this reason, starter cultures are rarely used for the second (mash) fermentation.

The long duration of the second fermentation step (6-12 months and 1-12 months for soya sauce and miso respectively) does offer opportunities for improvement by the application of such biotechnological principles as enzyme technology and/or starter improvement.

The starting material for sufu production is tofu, the pressed curd of soya milk (formed by acid or calcium coagulation of the soya proteins). Tofu contains 88% moisture, 6% protein and 3.5% oil. Tofu is cut into 3-cm cubes and soaked in 6% salt and 2.5% citric acid solution to prevent the outgrowth of contaminating bacteria. After draining, heating and surface drying, the cubes are inoculated with a fungal starter. Proteolytic and lipolytic cultures of *Mucor, Actinomucor* or *Rhizopus* are used. The fermentation process takes several days, depending on the organism used. The fermented tofu is then aged in various types of brines consisting of salted fermented rice mash, soya sauce moromi mash or fermented soya paste containing rice wine (10% ethanol). Flavours are formed during aging by chemical or enzymatic reaction between the hydrolytic products of fats and proteins of the tofu and the added ethanol. The product is a creamy, cheese-like substance, which is sterilised before packaging.

3.6.3 Tempeh production

Tempeh production is an example of a solid-substrate fermentation process in which a possible acid fermentation is followed by a fungal fermentation. Dehulled (ie with their thick-walled coats removed) soybeans are soaked in water (ca 30 minutes, 25 °C) and cooked before inoculation. Acidification occurs during soaking by lactic acid bacteria, preventing outgrowth of contaminants. After inoculation with *Rhizopus oligospora* a rapid outgrowth occurs, further controlling the activity of contaminants by substrate competition and even by production of antimicrobial substances. The inoculated beans are tightly packed in leaves to reduce air penetration. The most important biochemical changes are caused through lipolytic and proteolytic enzymes produced by the fungus. Additional positive effects are the increase in vitamin content (niacin, riboflavin, pantothenic acid and B_6), higher resistance of tempeh against auto-oxidation, elimination of the beany flavour and a softening of the beans by cellulolytic enzymes. The tempeh production process is interesting because a tasty, valuable protein-rich product is produced as a possible meat substitute from soybeans. The process is characterised by its simplicity, rapidity (20-24 h) and versatility. Cake-like tempeh products are also made from copra (tempeh bongkrek), lima beans (tempeh koro), bengkrek (tempeh bengkrek) and several grains such as wheat, oats, barley and rice.

In summary, Oriental food fermentation is not in the first instance directed at preservation (in contrast to acid food fermentation). Instead, in most processes, several traditional techniques such as drying, cooking or refrigeration are used to eliminate deterioration by contaminants and so ensure optimal fermentation processes.

Potentially, Oriental foods offer opportunities for a wider application of biotechnology. However, some problems need to be solved. Long fermentation times in rather salty conditions with mixed cultures are not in line with Western technological thinking. To be more acceptable, there would need to be a high-quality product generated in controlled processes with economical technologies. More know-how is therefore needed of the essential microbial and biochemical processes involved. After that, improvement by enzyme technology or by strain improvement might be possible.

3.7 Outlook and perspectives of fermented foods

In this chapter we have examined the ways in which micro-organisms contribute to the preservation and flavour of fermented foods. This relatively simple technology of preservation has led to an enormous variety of fermented foods worldwide. Besides this simple technology, renewed interest in food fermentations is caused by two other

SAQ 3.8	In column A are listed fermented foods. Match each product with the physiological activity in column B involved in its formation, and with the organism in column C which carries out that activity.

A	B	C
butter	cellulose degradation	*Lb. bulgaricus*
yoghurt	citrate utilisation	*Micrococcus spp.*
dry sausage	starch degradation	*Aspergillus oryzae*
koji	nitrate reduction	*L. lactis*
soya sauce	threonine degradation	*Rhizopus oligospora*
tempeh	alcohol formation	*Saccharomyces rouxii*

aspects: nutritional and therapeutic aspects of the foodstuffs produced. In many fermentations, improvement of nutritional properties of the food has been demonstrated. Examples are: synthesis of vitamins; predigestion of proteins and polysaccharides; elimination of anti-nutritional factors (such as phytates, glucosinolates and lectins); increased availability of minerals.

Lactic acid bacteria might also add some therapeutic activities to fermented foods. These are a reduction of serum cholesterol, possible anti-tumour activity, and antibacterial activity. Much research is necessary to prove these therapeutic actions conclusively.

Although a great deal of research is being carried out on the genetics of lactic acid bacteria (see Chapter 4) it might be some time before genetically improved strains produced by recombinant DNA technology are applied in practice. On the one hand this is due to a certain reluctance on the part of both producers and consumers to accept the use of such organisms in producing our food. This issue is discussed in more detail in Chapter 4 and Appendix 1. On the other hand, this chapter shows that, with few exceptions, the contribution and function of the lactic acid bacteria in the fermentation processes needs to be better defined before genetically improved strains may be constructed and find a place on the market.

Summary and objectives

In this chapter we have examined the use of fermentation to preserve a wide range of foods. Central to this discussion has been the use of lactic acid bacteria and a variety of food processes which rely on other fermentations. Eastern and western fermentation technologies have been contrasted. Now that you have completed this chapter you should be able to:

- explain how lactic acid contributes to the preservation of acid-fermented food;

- describe in outline the other measures taken in combination with acid formation to preserve acid-fermented foods;

- describe the biochemical pathways leading to the major products of homo- and hetero-lactic fermentation;

- describe activities (other than lactic acid formation) of micro-organisms important in the production of acid fermented foods;

- draw a flow diagram to show the industrial production of cheese, yoghurt and sauerkraut;

- outline the production of acid-fermented meat products;

- explain the role of micro-organisms in the production of koji, soya sauce and tempeh;

- draw a flow diagram showing the production of soya sauce and tempeh;

- relate particular micro-organisms used in food fermentations to particular products.

Starter cultures for cheese production

Starter cultures for cheese production

4.1 Introduction

The manufacture of cheese constitutes a major fermentation industry, particularly in the developed countries. For example, in 1988, 4.3 million tons of cheese with an estimated value of about US $20 billion, were produced within the European Community. The fermentation of milk to cheese is an important activity which requires a constant input of both new technology and fundamental research. The ancient craft of cheese-making has developed into a biotechnological process involving the fields of process technology, enzymology, microbiology and biochemistry.

Industrial cheese production requires a thorough understanding of the function of both lactic acid bacteria and rennet (chymosin) - see Chapters 3 and 5. These are involved in concentrating and preserving the milk protein (casein), lipids and other milk constituents as well as ripening of the cheese, and they determine, to a large extent, the cheese variety which is produced. In industrial cheese production, which has reached a high degree of perfection, the biological activity of the starter bacteria has always been the most difficult aspect to control. Therefore, there has been a need for the development of reliable starter cultures that may be used for the reproducible, large-scale production of cheese of a high quality. This chapter is aimed at describing the development and production of starter cultures for use in two popular varieties of worldwide significance, namely Gouda and Cheddar cheese.

4.2 The composition of starter cultures

mesophilic

Starter cultures for the production of Gouda and Cheddar cheese contain mesophilic lactic acid bacteria, which grow optimally at temperatures between 28°C and 32°C. Starters may be differentiated on the basis of:

- the species present;

- the number of strains present.

For Gouda and Cheddar, all starter cultures contain one or more strains of *Lactococcus lactis* subspecies *cremoris*, or *Lactococcus lactis* subspecies *lactis*. These organisms were previously named *Streptococcus cremoris* and *Streptococcus lactis*, respectively, and belonged to the serological group N of the genus *Streptococcus*. Many starter cultures also contain bacteria which are capable of utilising the citrate that is present in the milk. These include members of the genus *Leuconostoc* (mainly *Leuconostoc cremoris* and *Leuconostoc plaetis*) and the lactococcal variant, *L. lactis* subspecies *lactis* variant *diacetylactis*. These bacteria convert citrate into various compounds of which CO_2, important for eye formation in Gouda cheese, and the flavours diacetyl and acetic acid are the most significant (see Chapter 3). As a consequence, these species are known as

aromabacteria

aromabacteria .

The number of strains and their origin may vary widely, depending on the starter culture system used.

single-strain starters
Single-strain starters are monocultures of only one lactococcal strain. However, since a number of industrial properties of lactococci are unstable (see Section 4.6), the composition of a starter often changes dramatically after subculture.

multiple-strain starters
Multiple-strain starters are composed of several (up to six) different single strains and may include one or more aromabacteria.

mixed-strain starters
L-starters
D-starters
Mixed-strain starters contain a large (usually more than 50), undefined number of lactococci and aromabacteria. Their exact origin is not defined and they are not, as such, composed by man. These starters have been used for a long time in the dairy factories. During subculturing and subsequent cultivation no special precautions were taken to prevent contamination with bacteriophages. Depending on the nature of aromabacteria, mixed strain starters may be categorized as L-starters or D-starters if they contain only *Leuconostoc spp.* or *Lactoccocus lactis* subsp. *lactis* var. *diacetylactis* respectively as the sole citrate utilising bacteria. DL-starters contain both species and O-starters contain no aromabacteria.

| SAQ 4.1 | A traditional starter culture used for making cheese is known to contain over 60 difference strains of bacteria. Perhaps surprisingly, the only species present which could be properly called aromabacterium is a *Leuconostoc spp.* Which of the following terms can be applied to this starter culture?

1) Single-strain starter.

2) Multiple-strain starter.

3) Mixed-strain starter.

4) L-starter.

5) D-starter.

6) DL-starter.

7) O-starter.

8) Monoculture.

4.3 The industrial production of cheese

4.3.1 Gouda cheese

raw milk
Raw milk is collected, cooled and stored at low temperatures at the farm, until it is transported to the cheese factory where it is pumped into large tanks (up to 300 000 litres). Milk as delivered to the factories has a typical composition as is shown in Table 4.1.

Compound	Average % (w/w)	Compound	Average % (w/w)
lactose	4.6	minerals	0.7
fat	3.9	organic acids	0.17
protein	3.3	others	0.15
casein	2.6		

Table 4.1 The average composition of milk (g per 100g) used to make Gouda.

To compensate for variations in the milk composition, it is necessary to standardise the milk with respect to protein and fat content (for example fat content should be 3.2% for a full-fat cheese). This is necessary to reproducibly produce Gouda of the desired composition. This can normally be achieved by blending milk batches, skimming (centrifugation) to remove fat, or fat addition. The standardised milk is pasteurised (15 seconds at 72°C) and is subsequently cooled to 30°C before being transferred to the cheese vat or tank (a stainless steel vessel, with internal blades or stirrers, placed on a horizontal axis). At this stage of the process the milk is referred to as cheese milk.

pasteurisation

cheese milk

Π Why do you think that raw milk needs to be pasteurised as a preparation for cheese-making? (Write down at least one reason before reading on)

The main aim is to reduce the numbers of indigenous lactic acid and spoilage bacteria in the milk, so that the added starter culture can predominate. If this is not done, the indigenous bacteria may produce undesirable flavours and odours and so produce product of variable (and low) quality. In addition, pasteurisation kills any pathogenic organisms which may be present in the milk (and which may otherwise cause disease if eaten). An example of such an organism is *Listeria monocytogenes.*

non aseptic conditions

Because of the need for manufacture on a large scale and at low cost, it is impossible to operate the industrial cheese fermentations under completely aseptic conditions (such as in a closed fermenter system). As a consequence, some contamination will always occur, especially since the cheese milk is not sterilised and the equipment is used semi-continuously. The contamination that these conditions result in does not normally affect cheese quality. The number of contaminating spoilage organisms is usually low, and the starter culture predominates. However bacteriophage levels can build up to a point where starter activity is adversely affected. This problem will be discussed in more detail later on.

renneting process

syneresis

An L- or DL-(bulk) starter culture is added to the pasteurised milk in the vat or tank, 0.5 - 1.0% (v/v), to initiate fermentation. Other additions are rennet and calcium chloride to promote the renneting process (see Chapter 5), and sodium nitrate to prevent any butyric acid fermentation due to contamination by butyric acid bacteria in the milk. After 25 minutes at 30°C the gelled milk is cut or stirred to promote syneresis (exudation of the whey).

The syneresis is further promoted by heating the curd/whey mixture. In Gouda cheese production, this is done by the addition of hot water. The quantity of milk sugar (lactose) which is retained in the curd may be adjusted by varying the volume of water added.

cooking

The heating (cooking) temperature is usually 34°C. When the moisture content of the curd has been reduced sufficiently by stirring, the whey is drained off for 25 minutes. In a modern Gouda cheese factory, specialised equipment is used for drainage. This is usually a curd dispensing machine with twin-walled drainage tubes. The whey drains away through the perforation of the inner wall. The cylindrical curd blocks of young cheese are allowed to fall into prepared cheese moulds. After draining for 15 minutes

pressing

the cheese is pressed, resulting in a loss of more whey and the development of a closed rind. During pressing and the subsequent holding (storage), most of the lactic acid

holding

fermentation takes place. After holding, the cheeses are steeped in brine. Lactic acid

uncoupling

fermentation is uncoupled at this stage (that is lactic acid fermentation continues, although the organisms do not grow). Lactose conversion continues until all lactose has been fermented. This produces a pH value of about 5.4. After removal from the brine, the Gouda cheese is free from lactose. The cheese is then ripened by an appropriate period of storage.

Π The number of starter organisms in a bulk starter was $10^9 ml^{-1}$. This was inoculated into milk at a level of 1% (v/v). The cheese produced contains 10^9 starter organisms g^{-1}. When the milk curdled, bacteria were concentrated tenfold into the curds (assume none are lost in the whey). Using these data, can you work out how many generations (doublings) of the starter bacteria will have resulted? (Try to do this calculation before reading on).

The starter contained 10^9 organisms ml^{-1}. When inoculated into the milk at 1% v/v, the number of starter organisms would be 10^7 ml^{-1}. Upon curdling (and tenfold concentration) this number would become 10^8 organisms ml (g^{-1}) of cheese (assuming no loss). After 1 generation this number becomes 2×10^8 g^{-1}, after 2 generations 4×10^8 g^{-1}, after 3 generations 8×10^8 g^{-1}, after 4 generations 16×10^8 g^{-1}. Thus to reach levels of 1×10^9 g^{-1}, the organisms must have undergone just over 3 generations.

4.3.2 Cheddar cheese

difference between Gouda and Cheddar

The production of Cheddar cheese is in many aspects similar to that of Gouda cheese. The following differences are important for the production process. The Cheddar starter preferably does not contain aromabacteria since the production of CO_2 results in the formation of undesired eyes and/or slits in the cheese. Cooking is at 29.5°C for 40 minutes. Whey drainage and subsequent handling of the curd are performed in

Cheddaring

double-walled towers or on large conveyor systems. The so called 'Cheddaring' process involves heating (cooking at 38-40°C), cutting (milling) and pressing the curd. During this process most of the acidification takes place, pockets of gas are squeezed out and the curd gets its typical fibrous loose texture. Modern mechanised systems involve vacuum pressing of the curd pieces. The higher temperature of curd cooking (38-40°C)

uncoupling

uncouples growth from lactic acid production, controls the development of bacteriophage infection, and leads to development of important flavour components. Finally, the way in which salt is introduced in Cheddar differs significantly from the Gouda process, since salt is added directly to the milled curd. This prevents further growth of the starter bacteria and causes (with the high cooking temperature) the uncoupling of lactose fermentation from growth. Depending on the salt concentration, it may take some days or more than one week before all the lactose is fermented.

SAQ 4.2

The flow diagram below represents the production of Gouda cheese. Using the list below, fit the various outputs, inputs and operations into the diagram to complete it.

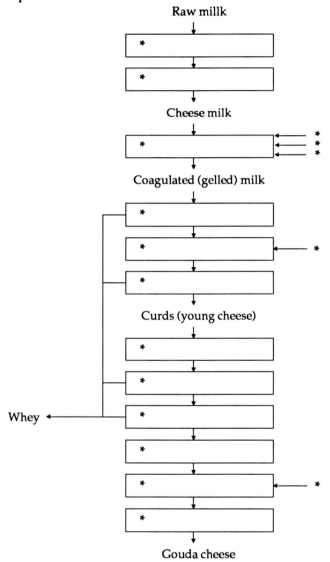

Pasteurisation/cooling
Starter culture (0.5-1% v/v)
Cooking (with stirring) (40 mins/34°C)
Draining (15 minutes)
Ripening
Standardisation (to 3.2% fat)
Pressing (75 mins)
Brining

Sodium nitrate
Cutting/stirring (20 mins)
Draining (25 mins)
Holding (75 mins)
Filling cheese moulds
Rennet
Incubation (25 mins/30°C)
Hot water
Brine

∏ Draw up a chart to contrast Cheddar and Gouda cheese manufacture. We would
     suggest you use the following format (example).

Process:	Gouda:	Cheddar:
Starter	contains aromabacteria	aromabacteria absent

∏ Write down two reasons why it is important that all the lactose in the cheese is
     fermented.

Obviously removal of lactose removes sweetness from the cheese. More importantly,
the lack of available sugar makes the cheese less prone to spoilage (in other words, it
aids preservation).

4.4 Activities of starter cultures

The important activities of starter bacteria in cheese production are:

• acid production (for promoting the exudation of whey and for preservation);

• formation of anaerobic conditions in curd;

• proteolysis and lipolysis;

• gas (CO_2) and aroma production.

rapid acid
production

The primary requirement of starter cultures is the production of lactic acid. Rapid and
reproducible lactic acid production is essential for industrial cheese production since it
has to be completed within a pre-fixed time schedule allowing computerised process
control. In addition, acid production is a main factor in the expulsion of moisture from
the cheese curd and determines the final moisture content of the cheese and hence its
texture and quality.

The two main mesophilic lactic acid bacteria, Lactococci and Leuconostocs, differ in the
mode of transport of lactose and the pathway utilised for lactose degradation as well as
in the amount and stereoisomeric form of lactate produced. Industrially-used lactococci
produce (E_h)L(+)- L-lactate and form only very small amounts of secondary metabolites
such as acetic acid and diacetyl. Leuconostocs produce the D(-)-isomer of lactate and
significant amounts of ethanol and (under conditions of high redox potential) acetate.
We have examined the energetics of lactose metabolism in these organisms in Chapter
2. Perhaps you should refresh your memory by re-reading the relevant section.

redox potential
changes in
cheese
manufacture

Another activity of starter bacteria, after inoculation in milk, is to metabolize the oxygen
present and to generate reducing compounds. The redox (E_h) potential decreases
rapidly to a value of about -200 mV, a value which increases again when the pH
decreases: after acidification the E_h of cheese is about -120 mV. A small number of
flavoprotein oxidase enzymes is responsible for the direct interaction of starter
lactococci with oxygen. Hydrogen peroxide (H_2O_2) or water are produced in these
reactions:

$$NADH + H^+ + O_2 \xrightarrow[\text{oxidase}]{\text{NADH H}_2\text{O}_2} NAD^+ + H_2O_2$$

$$2NADH + 2H^+ + O_2 \xrightarrow[\text{oxidase}]{\text{NADH H}_2\text{O}} 2NAD^+ + H_2O$$

lactoperoxidase

In some cultures, hydrogen peroxide accumulates to significant levels but is then further converted to H_2O by a peroxidase present in the starter bacteria. H_2O_2 can (together with the thiocyanate of the milk (1-15 ppm) and the milk peroxidase) form an inhibitory lactoperoxidase system for which most cheese starters are sensitive (see Section 3.7).

inhibition and killing of spoilage organisms

The preservation of cheese partly depends upon the microbiological activities of starter cultures controlling the growth of spoilage bacteria and pathogens. Inhibition or killing may result from the absence of oxygen, decrease in redox potential and pH and from the depletion of nutrients (eg lactose). In addition, inhibitory compounds such as H_2O_2, lactate and acetate are produced. Another class of antimicrobial inhibitors are bacteriocins. The polypeptide nisin, produced by *L. lactis* subsp *lactis* strains, is the best characterised bacteriocin and has wide application as a preservative in the dairy and food industry.

Because of its complex composition, milk is a satisfactory medium to support growth of the mesophilic lactic acid bacteria (which are nutritionally fastidious). However, various amino acids are either stimulating or essential for growth of starter bacteria. Since milk contains only small amounts of free amino acids or small peptides, starter bacteria need a proteolytic system that produces free amino acids from milk protein.

proteinase

The key enzyme in the proteolysis by lactococci is a proteinase located in the cell envelope that degrades the major accessible milk protein (casein) into peptides. These

peptidases

peptides are subsequently converted into free amino acids by several peptidases as shown below.

$$casein \xrightarrow{\text{proteinase}} peptides \xrightarrow{\text{peptidases}} amino\ acids$$

proteinase-deficient

Since the ability to produce proteinase is encoded by unstable plasmids in many lactococcal strains (see Section 4.6.1), there is a high incidence of the generation of proteinase-deficient (Prt') mutants. These are unable to grow in milk as pure cultures but may continue to grow in the presence of a (parental) Prt^+ strain that produces sufficient proteinase to allow the formation of peptides. This cross-feeding may result

genetic instability

in the accumulation of as many as 80% Prt' mutants in starter cultures.

The proteolytic system of starter bacteria is not only important for their growth but also for the hydrolysis of protein that takes places during cheese ripening. This proteolytic degradation is brought about by the action of starter enzymes and rennet (see Chapter 5). It results in the formation of small peptides and amino acids that contribute either directly, or as precursor for other compounds, to the flavour of the cheese. In some cases,

bitter-tasting peptides

this enzymatic hydrolysis results in the formation of bitter-tasting peptides (mainly from β-casein). Depending on the peptidase content and activities of the starter bacteria, these bitter peptides are converted into non-bitter peptides. It has been recognised for some time that starter strains may be differentiated on the basis of their capacity to produce non-bitter tasting cheeses.

Finally, the aromabacteria are able to ferment citric acid, which is metabolized to CO_2, acetic acid and C_4 compounds such as diacetyl or acetoin (see Chapter 3). Thus the important activities of starter cultures are: acid production, formation of anaerobic conditions, proteolysis and lipolysis, gas (CO_2) and aroma production.

4.5 Factors affecting starter activity

factors
influencing rate
of acid
production

Table 4.2 shows the factors affecting the main activity of the starter during cheese-making (ie the production of acid). It will be clear that it may be difficult to control all of these factors.

Main Factor	Example
1) Composition of the milk	level of free amino acids
	nucleotide concentration
	mineral composition
	glutathione concentration
	level of free fatty acids
2) Inhibitory compounds in the milk	milk immunoglobulins
	lactoperoxidase system
	oxygen
	contamination with detergents or disinfectants
	contamination with antibiotics
	contamination with bacteriocin-producing bacteria
3) Intrinsic factors of starter	physiological state
	genotype and phenotype
	composition of mixture (for mixed starters)
4) Manufacturing conditions	starter concentration (after inoculation)
	temperature
	salt content
	bacteriophage contamination

Table 4.2 Factors affecting the rate of acid production by the starter during cheese-making.

∏ Look through the Table 4.2 and see if you can think of a way in which each of these factors might be controlled.

Various modern developments, which are practiced already, minimise the impact of the majority of these factors. Improvements which result in a constant composition of the milk and prevent the presence of inhibitory compounds include: the spreading of the dates of calving and therefore the lactation periods; the modernisation of milk collection, storage and transport; the improvement of milk quality; the modern large-scale and semi-continuous production of cheese; the mixing of separate batches of milk (up to 300 000 kg) for cheese production. In addition, the effect of most of the factors inherent to the starter bacteria are minimised by the common practice of using standardised starter concentrates (see Section 4.8). Finally, the manufacturing conditions such as the temperature, salt content and starter concentration are kept constant in an automated cheese factory and so do not have a varying effect. In contrast, the effect of bacteriophages is more serious in large-scale and semi-continuous cheese production. We will focus on this aspect.

4.5.1 Bacteriophages

Lactococcal bacteriophages may have an immense impact on dairy fermentation processes. One reason for this is the rapid multiplication of virulent bacteriophages under optimal conditions.

∏ A bacteriophage has an infection cycle in cheese milk (the time between initial infection of the cell by one phage and the final release of many phages from the infected cell) of about 30 minutes. The burst size (the average number of phages produced from the infected cell) is 100. Cheese milk containing 1 bacteriophage ml^{-1} is inoculated with starter culture. Assuming that the bacteria susceptible to the bacteriophage remain in excess, what will the phage concentration be after 2 hours?

After 30 minutes the 1 phage will have infected 1 cell, and produced 100 phages (total = 10^2 ml^{-1}). After a further 30 minutes each of these phages will have produced a further 100 phages (total = 10^2 x 100 = 10^4 ml^{-1}). After a further 30 minutes the total will be 10^4 x 100 = 10^6 ml^{-1} and after the 2 hours, the total will be 10^6 x 100 = 10^8 ml^{-1}. By this time, therefore, more or less all of the susceptible bacteria will be infected. For several reasons the multiplication is less fast in practice. Nevertheless it is still immense.

Characteristics of bacteriophages infecting starter cultures

lytic temperate

Both lytic bacteriophages (capable of only a lytic life cycle in which the host cell is lysed) and temperate bacteriophages (capable of integrating into the genome of a host, without causing lysis) have been found in lactococci. A bacterium carrying a temperate bacteriophage is said to be lysogenic. Temperate bacteriophages are thus referred to as lysogenic bacteriophages. Although lytic bacteriophages are the main cause of problems in industrial fermentations, temperate bacteriophages also form a potential risk to a starter culture.

∏ If temperate bacteriophages integrate into the host cell chromosome without causing lysis, how can they form a potential risk in starter cultures? This is not an easy question so think about it for a little while before reading on.

prophage

The lysogenic bacterium (ie a bacterium containing a temperate bacteriophage) will replicate the incorporated phage genome (the prophage) along with its own chromosome. All the progeny of that cell will thus carry the prophage. Either spontaneously, or as a response to a certain stimulus, the prophage switches to a lytic cycle, lysing the host. The bacteriophages produced by this will undergo lysogeny with the same host bacterium, but may be lytic to other bacteria (in mixed starters).

lysogenic immunity

The fact that the host cell carrying an integrated prophage is protected against superinfection by related temperate bacteriophages (in a process called lysogenic immunity) is of little practical value. In addition, lytic mutants of a temperate bacteriophage may arise. The possibility of mutation means that the presence of bacteriophage, either lytic or temperate, forms a continuous threat since it may result in:

- lytic mutants;

- a more effective lytic bacteriophage;

- a lytic bacteriophage with an extended host range.

host range of
phages

The bacteriophages can be grouped according to their structure (as observed in the electron microscope), or by more sophisticated techniques such as nucleic acid homology. In this latter technique, use is made of the fact that closely related bacteriophages will have similar sequences of nucleotides in their nucleic acids. The host range of the various bacteriophages (that is the number of sensitive host strains) may differ considerably. Each bacteriophage however has a very specific host range.

Π   Do you think it would be possible to group the lactococcal bacteriophages on the basis of their host ranges? If your answer was yes, do you believe that such groupings would be reliable?

This is possible, and is a method often employed to group various phages. However it is not necessarily reliable, as the susceptibility/resistance of a bacterial host can alter (spontaneously) as a result of mutation. The mechanisms involved are discussed below.

Defence mechanisms against bacteriophages

cell wall
resistance

restriction-
modification

abortive
infection

lysogenic
immunity

Industrial strains of lactococci possess a variety of bacteriophage defence mechanisms which have contributed to their survival in the fermentation processes. These mechanisms are effective at various stages of the bacteriophage multiplication cycle. Cell wall resistance is usually associated with an alteration of the wall receptor sites and results in the failure of the bacteriophage to become attached to the cell wall. Intracellular protection may be realised through various mechanisms. One is restriction-modification which operates by a restriction enzyme degrading the foreign bacteriophage DNA. Another system is abortive infection, where the bacteriophage DNA is not converted into a functional bacteriophage particle. Finally, lysogenic immunity, which has been discussed above, may be regarded as a defence mechanism.

Bacteriophage-insensitive starters

P-starters

Traditionally, mixed-strain starters were propagated daily on farms and in factories without any controlled protection against contamination with phages. The continuous presence of bacteriophages has resulted in the natural selection of mixed-strain starters which are in equilibrium with their environment. In such conditions, only those that are resistant to the bacteriophages survive. These practice-derived (P)-starters usually have a high bacteriophage insensitivity (resistance). The mechanism underlying this insensitivity is not fully understood, but it includes cell wall resistance. The current idea is that due to a wide variety of strains with defence mechanisms against different bacteriophages, there is a large fraction of bacteriophage-insensitive strains within a P-starter.

The complexity of the P-starters is maintained by minimising subculturing (as this results in the dominance of fast-growing strains and loss of the natural diversity). In addition, P-starters usually contain a large number of phage particles (in the order of 10^7 ml^{-1}), which contribute to maintaining the insensitivity by preventing the growth of bacteriophage-sensitive mutants. The term P-starter is reserved for those mixed strains starters which have been obtained directly from practice and used after a minimum of transfers. Other mixed strain starters are also in use which do not fulfill these specifications.

starter failure
with single
strain starters

Upon infection of a P-starter by bacteriophage, the sensitive bacteria are destroyed and become replaced by insensitive strains, normally within a cultivation time of 22 hours at 20°C. The activity of the P-starter after recovery may be slightly diminished. However, its activity will be much higher than would be the case of a single strain attacked by phages. Since the fraction of insensitive mutants within a single-strain starter is rarely higher than 0.01%, bacteriophage infection usually leads to complete starter failure. The effect of a heavy phage infection on development of P-starters and single-strain starters is shown in Figure 4.1.

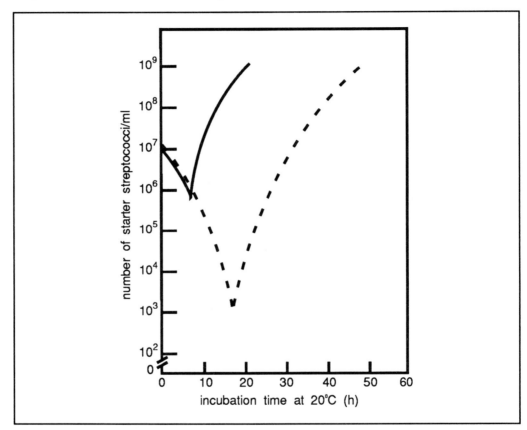

Figure 4.1 Effects of bacteriophage infection on P-starters and single-strain starters. Behaviour of a P-starter in milk after an attack by a disturbing phage (inoculation 1%, resistant strains 10%, solid line). Behaviour of a single-strain starter in milk after an attack by a homologous phage (inoculation 1%, resistant variants 0.01%, broken line).

4.6 The genetics of starter cultures

Most interest in the genetics of starter cultures originates from their application in dairy fermentations. In developing new and improved starter cultures, classic microbiological and genetic techniques have been important tools. These techniques include:

- screening single and mixed strains of lactic acid bacteria for desired properties;

- mutation of selected strains;

- the use of natural gene transfer systems to allow for the introduction of specific properties into selected strains.

In addition, genetic modification by recombinant-DNA technology (genetic engineering) is of great potential in improving starter cultures. We will now examine the various genetic modification procedures available for lactic starter cultures in more detail.

4.6.1 Plasmid biology

plasmids

Genetic material in bacteria can be arranged in two different ways: the major part as a long DNA molecule (the chromosomal DNA) encoding all important cellular functions; the minor part as smaller usually non-essential units of DNA (the extra-chromosomal, autonomously replicating plasmids). It has been well established that *L.lactis* and to a lesser extent, *Leuconostoc spp.* contain a variety of different plasmids. In some strains up to 12 plasmids per cell have been detected. The plasmid profiles may be obtained by size fractionation using agarose gel electrophoresis and are a valuable tool in distinguishing and identifying the strains used (see Figure 4.2).

plasmid-curing

It should be realised, however, that with a certain frequency (usually around 10^{-5} - 10^{-2} per generation) plasmids become spontaneously lost from a cell (a process called plasmid curing). Plasmid-free strains are not found in the industrially-used strains, but they may be constructed in the laboratory by repeatedly applying methods that increase the curing frequency, such as exposure to DNA intercalating dyes such as ethidium bromide or acriflavine, growth at sublethal temperatures, or protoplasting (removal of cell walls) and regeneration.

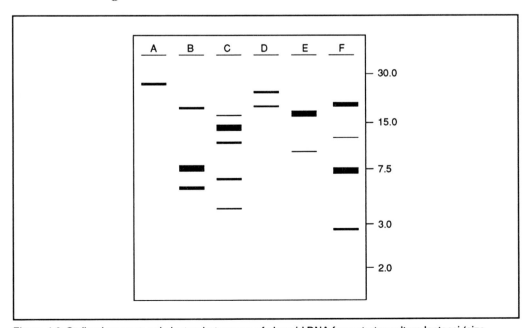

Figure 4.2 Stylised agarose gel electrophotograms of plasmid DNA from starter culture lactocci (size markers in megadaltons are indicated in the right hand lane - F Lane). Lanes A-E are plasmid profiles from 5 different starter cultures.

∏ A plasmid is spontaneously lost from an organism with a probability of 10^{-2} per generation. 100 bacteria containing the plasmid replicate through two generations. How many plasmid-free bacteria will there probably be?

At a frequency of 10^{-2} per generation, 1 in every 100 organisms will be plasmid-free. After 2 generations there will be 400 organisms. 8 of these are likely to be plasmid-free since if the frequency after 1 generation = 1 in 100 (ie 0.01 per generation), the frequency for 2 generations is 2 in 100.

∏ Which of the starter cultures illustrated in Figure 4.2 contains only a single plasmid type? Are all of the plasmids in culture C present in the same copy number (number of copies per cell)?

You should have spotted that culture A has only a single band. The density of the band reflects the amount of plasmid DNA. The results for channel C indicate that some of the plasmids are present in small amounts.

cryptic plasmids

functional plasmids

The majority of plasmids in the mesophilic lactic acid bacteria are cryptic (that is they encode for no known identifiable functions). However, it has been established that some plasmids are of primary importance in the application of dairy starters, since they encode key enzymes for important dairy functions (Table 4.3). Plasmids encoding for known functions are said to be functional. You will notice that the functional plasmids carry genes that modify metabolism or protect the host from phage infection.

Lactococcus spp.	*Leuconostoc spp.*
Lactose metabolism	Lactose metabolism
Proteinase production	Citrate metabolism
Citrate metabolism	Polymer production
Restriction-modification	
Bacteriophage adsorption	
Abortive bacteriophage infection	
Polymer production	

Table 4.3 Important plasmid-encoded functions in starter bacteria.

∏ Devise a scheme to show that a particular enzyme is plasmid-encoded.

The easiest way would be to cure the organism of the plasmid and see if the enzyme (or the enzyme function) disappears. By 'curing', we mean devise a scheme to cause loss of the plasmid. This can often be achieved by, for example, cultivating the organism at elevated temperatures. Alternatively the DNA sequence of the plasmid could be matched with the protein sequence of the enzyme.

The fact that these plasmids are inherently unstable (ie cultures can be easily 'cured') provides a rational explanation for the fact that important metabolic properties may be lost during subculturing of starter cultures. This effect was noted as long ago as the 1930s, long before the discovery of DNA as the carrier of genetic information.

Π   How do you think the problem of instability of important starter culture function on subculture could be minimised?

The answer is to minimise subculture as much as possible (obviously!). In practice this means that such starter 'cultures' are not maintained by repeated subculture. Instead, once a suitable organism or culture is isolated, a large number of stock cultures are prepared from it, and are stored at low temperatures (-70°C or in liquid nitrogen), or after lyophilisation (freeze-drying). For each production run, one of the stock cultures is used in inoculum preparation. This procedure is also applied in most other fermentation processes. Stock culture production is described later on.

stock cultures

culture storage

spontaneous curing

In some strains a relatively high spontaneous curing frequency is observed for some functional and cryptic plasmids. This explains why during repeated subculturing of a single strain, mutants with different plasmid profiles arise, so producing a heterogeneous mixture of strains.

4.6.2 Natural gene transfer systems

transduction

Natural gene transfer systems include transduction and conjugation. Transduction is the method by which host-cell genetic material is packaged in the head of a multiplying bacteriophage during infection and is transferred to the recipient during the re-infection process. It has been demonstrated in lactococci to promote transfer of both chromosomal and plasmid genes. Since the quantity of DNA that can be packaged in a particular bacteriophage is limited, it is probable that only part of a plasmid will be transduced. If this fragment contains the replication region of the plasmid, the result will be a replicating plasmid of reduced size. In the case where (parts of) the essential replication regions are not present, the resulting fragment may only be replicated when integrated into a replicating DNA molecule (that is the chromosome). Once integrated, the transduced gene may be stable, and will be replicated normally during chromosome replication. This way of integrating unstable plasmid-located genes in the chromosome via transduction has been used to stabilise lactose and proteinase genes in lactococci.

conjugation

Conjugation occurs when close cell-cell contact is made between donor and recipient cells. This can be promoted by centrifugation or filtration of cells, plating at high densities or even by immobilising cells in alginate beads. Conjugal plasmid transfer at relatively high frequencies (up to 10^{-1}) has been observed in lactococci and leuconostocs. Both intra- and inter-species conjugation of plasmids have been demonstrated.

4.6.3 Artificial gene-transfer systems

protoplast fusion, protoplast transformation, whole cell transformation, electro-transformation

The artificial systems applicable to mesophilic lactic acid bacteria are protoplast fusion and transformation. The latter system involves the genetic changes that are induced by the addition of exogenously added DNA. Three transformation systems are currently used, namely, protoplast transformation, whole cell transformation and electro-transformation. All have been used exclusively with plasmid DNA. We will consider each of these in more detail.

Protoplast fusion

lysozyme

treatment

osmotically
stabilised

Since lactic acid bacteria are Gram-positive they may be converted into protoplasts by removal of their cell wall using lytic enzymes such as lysozyme. To prevent cell lysis, this treatment has to be performed in the presence of an osmotic stabiliser such as 0.5 M sucrose (in a hypotonic medium the protoplasts would swell and burst). The osmotically-stabilised protoplast may be regenerated into viable walled cells by growth on special regeneration media. Fusion of mixtures of protoplasts of different genetic types can be induced by the use of agents such as polyethylene glycol. This results in the initial formation of a diploid cell consisting of donor and recipient genomes, from which haploid cells will segregate on cell division. In the regenerated offspring, recombination between the different chromosomes or plasmids may have occurred thereby producing new strains.

Protoplast transformation and electro-transformation

transformation
with
protoplasts and
whole cells

In the early 1980s, protoplast transformation was the only transformation system that was applicable to some lactococcal strains. It was based on the induction of DNA uptake in protoplasts by the addition of polyethylene glycol (as in protoplast fusion) and relied on an efficient method for regenerating the recipient protoplasts. Another transformation system that was also applied to a few strains was based on the addition of polyethylene glycol and DNA to whole cells. Unfortunately, both methods were not applicable to industrially-used strains. To overcome this problem, electro-transformation was introduced, based on the formation of temporary permeability in whole cells after application of a high-voltage electrical field. This

electroporation

simple method, also known as electroporation, is applicable to a large number of bacterial species and includes most strains of *L. lactis* and *Leuconostoc spp.* In addition, the frequency of transformation using this method is relatively high (up to 10^6 transformants per μg plasmid DNA). At present, electro-transformation is the method of choice for introducing natural or modified plasmid DNA into starter bacteria.

4.6.4 Recombinant selection systems

Since starter bacteria are to be used in food fermentations and are intended for consumption, all applied genetic selection methods should be food-grade. This applies to the methods for selecting recombinant cells in conjugation, transduction, protoplast fusion and transformation.

∏ Write down a list of commonly used markers used for the selection of microbial recombinants. Now explain why these are not applicable to lactic starter cultures.

We have assumed that you have written down antibiotic resistance since these are the genetic markers most easily applied to recombinant selection. The insertion of such genes into organisms destined for consumption is obviously undesirable, especially if the gene is plasmid-borne (transmissible). You might have written down various nutritional markers (eg a requirement for an amino acid). Auxotrophy of this sort is a commonly-used selection marker, but, although food-grade, is undesirable as the organism would only grow well in milk or cheese if the required growth factor was present in high concentrations.

Food-grade selection markers currently used include the ability to ferment various sugars, resistance to bacteriophages and the production of and/or immunity to bacteriocins.

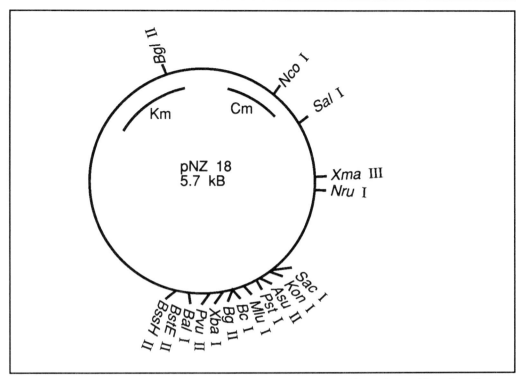

Figure 4.3 Physical map of a wide host-range vector pNZ18. This plasmid contains a wide variety of restriction enzyme sites and two antibiotic resistance genes (Cm = chloramphenicol; Km = kanamycin). This plasmid also contains a pSH71 replicon (not shown).

4.6.5 Cloning, expression and secretion vectors

A wide variety of cloning vectors have been developed, that should be useful in the genetic engineering of mesophilic lactic acid bacteria. These have been based on cryptic and functional plasmids as well as replicons from other Gram-positive bacteria. Of particular interest is a family of plasmid vectors based on a small (2060 nucleotides), cryptic plasmid of *L.lactis* that is widely distributed in industrially-used lactococci. These vectors, from which a representative is shown in Figure 4.3, replicate in all lactic acid bacteria investigated so far and are also able to replicate in the genetic model bacteria *Escherichia coli* and *Bacillus subtilis*. This property, and the presence of various unique restriction sites make these plasmids good candidates for use as cloning vectors. Improvements which can be made to make these plasmids more suitable as cloning vectors for food-grade use include:

cloning vectors

expression vectors

• the introduction of efficient signals for transcription and translation to make so-called expression vectors. These have been used to express a variety of homologous and heterologous genes in starter bacteria;

secretion vectors

• the introduction of signals promoting the secretion of the produced proteins to make so-called secretion vectors;

• replacement of antibiotic resistance selection markers with food-grade selection markers.

4.6.6 The prospects for applying genetically improved starters

A great advantage of the use of modern genetic techniques is the accuracy and high predictability of the nature of the genetic changes. This is in contrast to the random and frequently multiple changes obtained by mutagenesis.

∏ Do you think that mixed-strain starters are good candidates for genetic improvement?

These starters are complex, undefined and of variable composition. In such a situation, the effect (on manufacturing) of genetic improvement in one or more strains is hard to predict and would not necessarily last very long. Genetic improvement of single-strain or multiple-strain starters would be more likely to produce predictable and stable effects.

influence of legislation and consumers on choice of strategy

The different categories of gene manipulation methods that we examined in the previous section have very different potentials, as regards immediate application in the dairy industry. This is mainly because of legislation and the attitude of consumers, both of which, in turn, influence the attitude of producers. Let us examine these aspects in more detail.

wide application of natural systems

The first category (that is improved starters obtained by applying natural gene transfer systems) is very important, since the plasmids that encode important industrial properties are naturally transferable. If use is also made of food-grade selection systems, no limitations on application are to be expected, since the natural genetic barriers between species are not crossed. At present, such genetically improved starters are marketed worldwide (see Section 4.7.6).

difference between homologous and hetero-logous gene transfer in terms of legislation and acceptance

The second category consists of starter strains that have been constructed employing artificial gene transfer systems. When use is make of homologous and naturally occurring plasmids that are transferred using food-grade selection systems, no limitation on their application may be expected. Strictly speaking, such improved strains are genetically modified, but it has been generally accepted that the nature of the genetic change and not the method by which a change is obtained should be the criterion on which to assess the applicability of genetically modified organisms.

When, however, artificial gene transfer systems are used to clone and express heterologous genes, the situation is not as simple. By heterologous genes we mean non-homologous genetic material that would not recombine under natural conditions with the DNA of the host organism. Not only the construction but also the application of these genetically modified starter bacteria are subject to approval of legislative bodies. This is due to the fact that viable starter bacteria are present in cheese, even after long times of ripening (although their number may decrease more than 10^5-fold). This means that the genetically modified starters are not contained. At the time of writing, there have been no precedents for the uncontained use of such genetically modified micro-organisms in foods. Decision taking on this issue by legislative bodies is complicated, and acceptance by the consumer may be limited. Therefore, it is not foreseen that the uncontained use of genetically modified starter bacteria will become common practice in the near future. However, where the genetic changes include homologous or other 'safe' genes that have been modified using food-grade techniques, it may be expected that those initial barriers may be overcome and well-defined improved starter strains will be used for the production of cheese and other fermented foods. Some legislative aspects covering the use of genetically engineered organisms or their products are covered in Appendix 1 and in Chapter 5.

SAQ 4.3	A naturally occurring plasmid of *Lactococcus lactis* carries a gene for proteinase production. The proteinase is active at 35°C and at room (ripening) temperature. The plasmid is isolated and the proteinase gene is mutated *in vitro* in such a way that it is inoperative at 38°C (ie it is not active at the cooking temperature in Cheddar cheese-making) but remains fully active at room temperature. The mutated plasmid DNA is amplified and re-introduced by electroporation of protoplasts of a host derivative from which the proteinase plasmid has been cured by protoplasting and regeneration. A recombinant organism results, which is proteolytic at room temperature but not at 38°C.

1) Which of the following terms can be correctly applied to the gene-transfer system used?

a) Conjugation.

b) Transformation.

c) Protoplast fusion.

d) Transduction.

2) Would the process described in the question include gene-transfer of a heterologous or homologous gene?

4.7 Starter culture systems

Before we explore specific 'systems' of cheese production, let us remind ourselves of some of the objectives of the cheese manufacturer which will influence the selection and propagation procedures adopted for starter cultures.

The manufacturer will aim to select starters and carry out propagation procedures which:

- reduce the likelihood of contamination by bacteriophages;

- do not produce bitter off-flavours;

- reduce the ripening time required. This reduces storage and labour costs.

For historical reasons various strategies for selecting starters have been adopted in different countries. We will now examine the main ones.

4.7.1 The Dutch system

Some 15 years ago, mesophilic mixed-strain starters were collected in The Netherlands from factories where they had been transferred over a long period of time without precautions against bacteriophage contamination.

starter
concentrates

Since then, a large collection of such P-starters has been maintained at NIZO (The Netherlands Institute for Dairy Research). These have been maintained at neutral pH at -196°C (in liquid nitrogen), using 6-7% lactose as a cryoprotective agent. Those that appeared to be the best suited for cheese-making (those producing the fastest acidification rates and the finest cheese flavour) have been selected for producing concentrated starter cultures. They are used in this form for preparing the bulk starter (see Section 4.8). Transfer of the starter is kept at a minimum, both during the production of concentrated stock cultures and at the cheese production stage. In the latter case, this is achieved by using each time an identical frozen stock concentrated culture for preparation of the bulk starter. This ensures that starters with identical properties are employed.

bulk starter

Although P-starters will seldom fail completely when cultivated in this way, the activity of these starters still fluctuates during cheese-making due to bacteriophages. The various precautions and the equipment that are used to obtain bulk starters free from disturbing bacteriophages and to prevent bacteriophage accumulation during prolonged cheese-making are described in Section 4.9. Using this system it appears possible to use in a factory the same starter for Gouda cheese production year after year.

The system cannot only be applied to D- or DL-starters but also to O-starters suitable for making Cheddar cheese.

SAQ 4.4

Which of the completions to the statement is/are true?

To obtain an O-starter from a L-starter:

1) aromabacteria are removed;

2) aromabacteria are added;

3) *Leuconostoc* sp. are removed;

4) *Leuconostoc* sp. are added;

5) L-starters are added.

In practice, O-starters have been developed from D-starters by transferring D-starters a few times in milk with a low manganese content (derived from cows starting lactation in the spring) at an incubation temperature of 25-30°C. Under these conditions, the development of *Leuconostoc spp.* is so slow that they are lost when low inoculum levels (0.01% v/v) are used for the transfers.

4.7.2 The New Zealand system

single strains

Starters consisting of selected single strains were first introduced in New Zealand. Later, the procedure of rotating pairs of single strains was adopted. In this procedure up to 8 bacteriophage-unrelated strains, designated A to H, were initially selected for cheese production. On day one pair A and B, on day two pair C and D, on day three pair E and F and day four pair G and H were used as starter. On the fifth day the sequence of rotation restarted. Later, many new strains were introduced, and a longer rotation period of up to 14 days was practiced. The purpose of rotating starters is to prevent in factory accumulation and contamination of bacteriophages that are specific for the

rotating pairs

starter bacteria in use on a certain day. Although some success was obtained with rotating pairs of single strains, the system failed in larger mechanised cheese factories with a prolonged cheese-making period. The most important step to improve the situation was taken in the 1970s and involved a careful selection of specific strains. The major property used in this selection was resistance to the wide variety of bacteriophages present in cheese factories. Other properties included absence of temperate bacteriophages that were able to propagate lytically on other selected strains, rapid acid production rates and characteristic flavour production. The first of a newly isolated pair of single strains was used commercially in 1978.

complementary properties of starter culture pairs

The reason why a pair of strains is used needs some explanation. One strain of the pair is a temperature-sensitive strain, that stops growing at the cooking temperature of 38°C applied in Cheddar cheese-making. The other strain is a temperature-insensitive one, that continues developing during cheese-making. In the absence of bacteriophage contamination, the acid production rate is determined largely by the temperature-insensitive starter strain. In addition, the total number of starter cells in Cheddar cheese affects the degree of proteolysis and consequently the flavour of the final product. Since it is possible to vary the ratio between the two strains, both the acidification rate and flavour development can also be altered.

∏ Do you think that the temperature-sensitive or temperature-insensitive strains of the starter culture would be the most prone to bacteriophage infection? (Make a choice before reading on).

The temperature-sensitive organism will grow only throughout the initial fermentation phase (which lasts for about 40 minutes for Cheddar cheese (see Section 4.3.2). The temperature-insensitive organism will grow through the same period, as well as subsequent processing (cooking, Cheddaring and pressing) and even maturation. This longer growth period allows greater opportunity for phage build-up and infection.

improved hygiene

During the last decade, many cheese plants in New Zealand have increased their hygienic standards and taken other precautions (see Section 4.9), which enable continuous use of single pairs of strains. If a disturbing bacteriophage is encountered nowadays it develops very slowly under industrial conditions. An important aspect of

reduced bacteriophage diversification

this single-pair system is the fact that the number of potential hosts for contaminating bacteriophages is reduced. As a consequence, there is less opportunity for different bacteriophages to recombine or for mutation to occur. This limits the possibility of phages being able to increase their virulence or extend their host range.

The starters which have been used in New Zealand since the 1970s are produced by the New Zealand Dairy Research Institute and are distributed now as frozen concentrates to the various cooperative industries that use them.

4.7.3 The whey-derived system

The system of whey-derived starters has been developed in New Zealand and Australia, and is similar to the system described above. However, the selection of the appropriate strains is less strict and the strains do not always have a markedly increased resistance to bacteriophages.

The concept involves the use of a single strain and its replacement by resistant derivatives as soon as bacteriophage appears.

∏ How do you think phage-resistant starters can be derived from a phage-sensitive one?

challenged

The sensitive host is grown up (in reconstituted sterilised skimmed milk) and exposed to (challenged with) the phage (or phages) causing the problem. The phage culture can be obtained by passing cheese whey from affected batches through a 0.45μm or 0.22μm pore-size filter. This will remove bacterial contaminants but not the phage particles. Organisms which continue to grow on subculture in these conditions will be phage-resistant. Individual organisms can be isolated by plating for single colonies on an agar plate which incorporates suitable nutrients and the phage(s).

A special soft-agar plating medium is then used to differentiate between fast and slow milk-coagulating mutants (those which produce proteinase and those which do not, respectively) which are bacteriophage-resistant. The capacity of the isolated strains to produce sufficient activity and a fine-flavoured product is determined. Finally, it is verified that the basis of bacteriophage resistance is a receptor site mutation and not a result of immunity due to lysogeny.

∏ Devise a scheme to verify that the basis of the resistance of a culture to bacteriophages (compared to the parent culture from which it has been derived) has been an alteration in the phage receptor site.

plaque assay

bacteriophage absorption test

The test used is a bacteriophage adsorption test. A suspension of the bacteriophage can be produced by infecting the parent culture. The number of phage particles can then be counted by plaque assay. A standardised dense suspension of the parent and the 'derived' bacterial cells are each infected with a given number of phage particles. After allowing time for phage adsorption, the bacterial cells are removed (by centrifugation or filtration), and the number of phage particles remaining in the liquid are counted. This procedure indicates how many phage particles are adsorbed onto the bacterial cells.

daily check for bacteriophages

It is possible that the resistant starter strains that are obtained by such challenges will not be resistant to bacteriophages that emerge in the factory at a later date. If this occurs, new replacement strains have to be isolated using the described challenge approach. In the factory, a daily check is made for the presence of bacteriophage by testing the capacity of the starter to produce acid in a milk medium. In addition, samples of the wheys are saved and are tested for the presence of bacteriophage in a testing centre.

bacteriophage-insensitive mutants

It is evident that this system, which is also known as the bacteriophage-insensitive mutants (BIMs) system, is labour intensive and time consuming (and thus expensive) as so much routine testing has to be performed. Nevertheless, whey-derived starters are used for the majority of cheese produced in Australia and form part of the multiple-strain starter system, described below.

4.7.4 The multiple-strain starter system

Research in the USA, Ireland and recently also Germany, has resulted in the commercial application of the multiple-strain starter system. Initially, cultures of six strains were blended but later a lower number of strains (up to three) has been used. In general, cultures of the different strains are propagated separately. The methods of selecting the ideal strains vary throughout the world, and are continuously evaluated and improved as new selection strategies emerge. In some places, strains showing insensitivity to bacteriophages present in the cheese factory wheys have been used. In most cases the essential steps present in the whey-derived system, including the use of BIMs, are used to isolate the final strains.

cultures grown separately

In contrast to the New Zealand system, no rotation of the cultures is applied. Substantial savings have been reported when compared to the previously used mixed-strain starters. Although in most cases only Cheddar cheese is produced with the multiple-strain starter system, it is also feasible to apply the selection strategies to citrate-fermenting lactococcal strains in order to allow for the production of starters for Gouda and other cheese.

4.7.5 The proteinase-deficient (Prt⁻) starter system

Recently, there has been interest in using only proteinase-deficient strains as starters for cheese-making. It has been claimed that these starters are insensitive to bacteriophages. Depletion of available essential nitrogen compounds is the reason for very poor development of Prt⁻ mutants in milk (see Section 4.4) Therefore, Prt⁻ starters are produced in milk containing 0.1% yeast extract and a large amount (8% v/v) of the starter is added to the cheese milk to allow for sufficient acid production in the young cheese. Under these conditions, the Prt⁻ cells indeed appear to be insensitive to bacteriophages. However, if growth stimulants are added, the strains become bacteriophage-sensitive, indicating that it is not the genetic properties of the Prt⁻ strains, but rather the reduced growth rates, which cause bacteriophage-insensitivity. The simple explanation for this phenomenon is that for an efficient multiplication of bacteriophages actively growing cells are required.

poor growth of Prt cells

relative insensitivity to bacteriophage

SAQ 4.5

Choose the correct completion to the following statement.

Prt⁻ starters alone cannot be used for the production of **matured** cheeses because:

1) they do not secrete enough proteases to enable them to grow to such an extent that enough acid is produced to cause the coagulation of the milk;

2) they do not secrete enough proteases to provide themselves with essential nutrients they require for lactic acid production;

3) they do not secrete enough proteases to assist the other enzymes present in the ripening process;

4) the proteases they secrete are not able to cause the ripening process.

The drawbacks to the use of the Prt⁻ starter system are many. Firstly, there are the high volumes of inocula that are needed to start fermentations. In addition, carry-over of growth stimulants (such as yeast extract) takes place, which could influence the development of secondary microflora in the cheese.

SAQ 4.6

Choose the correct completion to the following statement.

A Prt⁻ mutant which arises spontaneously in a culture of Prt⁺ cells in milk cannot outgrow (completely take over) the parental strain because:

1) in the absence of Prt⁺ organisms Prt⁻ organisms cannot grow well;

2) the Prt⁻ gene is plasmid borne and therefore unstable;

3) Prt⁻ organisms have lost a plasmid;

4) Prt⁺ organisms can grow well especially in the presence of Prt⁻ organisms.

4.7.6 Genetically improved starter systems

One of the main targets in improving starter cultures is increasing the insensitivity to bacteriophages of the *L.lactis* strains used. Conjugation has been used for this as a natural and efficient transfer system and use is made of the common bacteriophage-insensitivity plasmids.

importance of
plasmid
pTR2030

In studying defence mechanisms against bacteriophages in strain *L.lactis* subsp. *lactis* ME2, plasmid-encoded resistance appeared to be encoded by plasmid pTR2030. Strain ME2 is a prototype bacteriophage-insensitive strain that is an important component of starter cultures used for Cheddar production in the USA. Plasmid pTR2030 encodes for a bacteriophage defence mechanism that results in abortive infection. pTR2030 may be transferred by conjugation to industrially useful *L.lactis* subsp. *cremoris* and *L.lactis* subsp. *lactis* strains. This has been achieved using high-efficiency conjugal donor strain that was made Lac⁻ (ie unable to metabolism lactose) by curing of the lactose plasmid. Transfer of pTR2030 to commercial lactose-utilising strains was thus realised using food-grade selection techniques.

Π Which of the following media would select for recombinant production strains arising from a phage-sensitive recipient after conjugation by Lac⁻ donor carrying pTR2030? Assume the basic medium provides suitable nitrogen sources, vitamins and minerals but lacks carbohydrate (and other sources of carbon and energy). 1) Basic medium; 2) Basic medium and lactose; 3) Basic medium and phage; 4) Basic medium and lactose and phage

The appropriate medium is 4). No organism (recipient, donor or recombinant) would grow in medium 1) or 3) as they do not have a source of carbohydrate. 2) would allow the growth of both the recipient strain and the recombinant. 4) would not allow growth of the Lac⁻ donor as it cannot use lactose and would not allow growth of the recipient which is phage sensitive. It would allow growth of the recombinant, which is the only type of organism which would be able to use lactose and have phage resistance.

Π   In the conjugation procedures described above, the process selects against the presence of the donor organism carrying pTR2030 (this is called counter selection). Is this necessary?

Yes! The donor strain is not wanted. Although it is phage resistant, it has properties that limits its industrial use.

introduction of
new strains
into the market

The presence of pTR2030 in the recipient strains was verified by isolating their plasmids and hybridisation with a synthetic oligonucleotide that had been designed on the basis of a partial sequence of pTR2030. What this means is that short single-stranded DNA fragments were made using chemical synthesis that have a nucleotide sequence identical to part of pTR2030. As a consequence these synthetic oligonucleotides are capable of hybridising (forming double helices) with pTR2030 DNA that had previously been denatured to form single strands. The synthetic probes can therefore be used to specifically detect the presence of pTR2030 DNA. Usually the synthetic oligonucleotides are labelled in some way (eg made radioactive) so that they can be detected. The recombinant strains were able to ferment lactose rapidly. Further testing showed stable maintenance of pTR2030 and the ability to produce proteinase. In the USA, introduction of the improved strains into cheese plants was initially carried out without indicating the nature of their improvement. Sales of the improved strains compared favourably with that of the traditional starter cultures. At a later stage, the cultures were advertised as being improved by genetic techniques and this resulted in a further increase of sales.

failure of
pTR2030 to
inhibit all
bacteriophages

Not all bacteriophages have been found to be inhibited by the presence of pTR2030, indicating the need for other bacteriophage defence mechanisms to be present. However, the phages that are the most common cause of starter failure are completely inhibited by pTR2030. After commercial use for over a year, one bacteriophage has evolved virulence against a pTR2030-containing starter culture. Nevertheless, the initial commercial success shows that starter cultures which are rationally improved by food-grade and acceptable genetic techniques, do have a valuable role in improvement of starter performance.

SAQ 4.7	In column A are listed various starter selection/production systems. For each starter system, match it with the appropriate starter type from those listed in column B and the appropriate mode of use from those listed in column C.

	A		B		C
i)	New Zealand system (before 1970)	a)	3-4 mixed strains	1)	No rotation (no action taken upon phage infection)
ii)	Whey-derived system	b)	Single strain	2)	No rotation (derive resistant strains upon phage infection)
iii)	Multiple-strain starter sytem	c)	Paired strains (phage resistant)	3)	Rotation

4.8 The production and use of starter cultures

Traditionally, in the cheese production plant, starter cultures (which are mixed-strain) were maintained by daily subculture in heat-treated milk. The culture maintained in this way is referred to as the mother culture. A portion of the mother culture was taken as required and inoculated into a larger volume of low-count milk. After incubation, this larger-volume culture, called a bulk starter was used to prepare more bulk starters, or used to inoculate cheese milk.

mother culture

bulk starter

SAQ 4.8

Which of the following are associated with traditional mixed-strain starter production compared to other types of starter?

1) Higher probability of loss of proteolytic activity due to loss of plasmids.

2) Higher cost due to the labour involved.

3) Higher probability of loss of ability to ferment lactose due to loss of plasmids.

4) Higher probability of contamination by bacteriophages.

5) Higher probability of change in strain composition.

---

stock cultures

The problems with the traditional production of starter cultures were partly overcome by renewing at intervals the mother culture used. The culture used for this was often provided by a dairy institute or a culture production company. Such a culture would be prepared from a bank of frozen or lyophilised cultures, called stock cultures.

preservation of cultures

With the ready availability of deep freezers and freeze driers, many cheese production plants carried out their own preservation of cultures. Starters could be preserved at any stage of production. Cells from mother cultures or bulk starters could be concentrated (by centrifugation or vacuum drying) and frozen or dried. Under the appropriate conditions, the activity of starter bacteria may be well preserved. In the case of deep freezing, these conditions include a final pH between 6.0 and 7.0, the presence of milk constituents that act as cryoprotectants (in particular lactose) and a storage temperature below -35°C. Starters can easily be preserved by inoculating cooled-steamed skimmed milk with 1% (v/v) of the starter and freezing immediately thereafter. Twenty four hours before the starter is needed, the frozen inoculated milk is thawed and incubated at 20°C. This produces an active culture ready for use.

∏ Compared to deep freezing, what do you think the advantages and disadvantages of culture preservation by freeze drying might be?

In freeze drying, the culture volume and weight are reduced and subsequent storage at ambient temperatures is possible. These make storage and/or transportation much easier and cheaper. However, freeze drying can lead to prolonged lag phases when the culture is reconstituted. Modern methods of freeze drying have largely overcome this problem.

During the last decade or so, both deep freezing and freeze drying have proven to be effective ways of reducing the risk of contamination and culture changes. These preservation methods have also allowed the centralised production of active starter concentrates and their distribution to cheese factories where they may also be stored until use. Such starter concentrates are used in the factory to inoculate heated milk to produce bulk starters, or for direct inoculation of cheese milk. Concentrated starters of this type used for direct inoculation of cheese milk are referred to as direct-to-vat-set (DVS) concentrates.

direct-to-vat-set
concentrates

Culture companies and dairy institutes now prepare concentrates of starter bacteria containing at least 3-5 x 10^{10} cells ml^{-1}. In The Netherlands, the majority of cheese is produced with mixed-strain cultures according to the Dutch starter culture system. Concentrates of these starters are produced on a fairly large scale according to the manufacture scheme such as that shown in Figure 4.4. Other institutions prepare these products in a similar way. At NIZO, the concentrates are placed in polystyrene beakers which are packed in boxes and blast chilled in an air stream with a temperature of -45°C. The frozen concentrates are distributed to the factories in insulated containers and stored in the factories at temperatures below -35°C. One batch consists of a great number of beakers. Each day one beaker from the batch is used for inoculating the bulk starter milk, thus ensuring the reproducibility of the inoculum.

∏ Mark onto Figure 4.4 the stages which could be described as a) medium preparation b) down-stream processing (do this before reading on).

We regard medium formulation and sterilisation as medium preparation, while all of the remaining steps except incubation can be regarded as down-stream processing.

DVS culture
costs when
high inocula
densities used

DVS cultures containing at least 5 x 10^{11} cells g^{-1} (or ml^{-1}) are commercially available in the form of frozen pellets or as free-flowing freeze-dried powders that disperse almost immediately upon addition to the milk. DVS concentrates differ from the traditional (acid) starter cultures in milk. When applied to cheese-making, the manufacturing procedure has to be adopted to compensate for the higher pH of milk. Another aspect is the cost of DVS starters. Especially when high inoculation percentages of starter are needed to start the fermentation process, the application of DVS concentrates will be rather expensive. Nevertheless, the DVS concentrates have considerable potential in fermentations which traditionally do not use a high inoculation percentage of the bulk starter, such as the thermophilic manufacture of Italian and Swiss-type cheeses.

∏ Make a list of the advantages to the cheese manufacturer of using DVS starters to the cheese. When you have done this, see if your response coincides with our ideas described below.

The DVS starter system avoids the need for the production of mother cultures and/or bulk starters on site. This is more convenient, it may save money and it reduces the risk of phage contamination. All that needs to be done is to empty the contents of the beaker/can/sachet into the cheese milk.

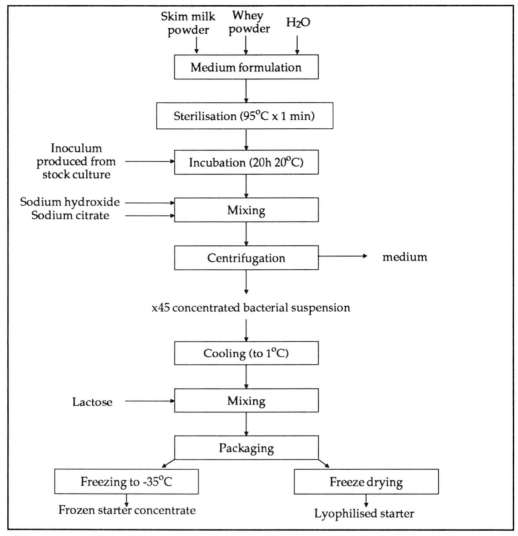

Figure 4.4 A flow diagram for the preparation of starter culture concentrates.

4.9 The prevention of bacteriophage contamination

4.9.1 Prevention of bacteriophage development during starter production

Low numbers of bacteriophages in milk inoculated for the production of bulk starters can be disastrous, since there is ample opportunity for them to multiply. There are two ways of dealing with this problem:

- preventing bacteriophage contamination from occurring;

- inhibition of the multiplication of bacteriophages in the starter milk.

Prevention of bacteriophage contamination

∏ Mentally, make a list of the possible sources of phage contamination in starter cultures.

We can think of the following sources:

1) the milk medium/ingredients used in cultivating the starter;

2) the vessel in which the starter is grown;

3) the equipment used to process the culture (centrifuges, driers);

4) the inoculum used;

5) the environment (for example aerosols in the atmosphere)

∏ Look through the list of sources above and think of ways in which contamination can be prevented.

1) The media used in propagating starters can be sterilised by heat. In practice pasteurisation (at 95°C for 15-30 minutes) is enough to kill bacteriophages present, and such treatments are used routinely.

2), 3) Vessels and equipment used can be sterilised/pasteurised by steam or hot water.

4) If you do not have any ideas on how infection in inocula for starters is minimised, re-read this chapter!

5) The best way to prevent external contaminants gaining entry to the culture vessel is to pressurise it (so that if there are any leaks, things leak *out* not *in*). This is especially important when the culture vessel is cooling down after steam pasteurisation. In practice high efficiency (HEPA) filters can be used to filter the air introduced into culture vessels. Such filters are highly effective in removing phage particles from air streams. Air over-pressure in the culture vessel headspace is maintained throughout the propagation period.

In The Netherlands, the fermenters used are additionally supplied with a specially designed inoculation beaker to allow for aseptical inoculation with the NIZO starter concentrate. Figure 4.5 shows a schematic representation of a bulk starter fermenter that is widely used in The Netherlands.

The use of phage-inhibitory media (PIM)

requirement of Ca^{2+} for phage adsorption

In common with many other bacteriophages, lactococcal phages require Ca^{2+} ions for effective adsorption to host cell receptors (and thus for infection). Phage-inhibitory media (PIM) have phosphate, citrate or other ion chelators added to them to sequester (effectively remove) Ca^{2+} ions. The use of PIM is extensive in the USA, and to a lesser extent elsewhere. It is often used in combination with the methods we discussed in the previous section. However the method cannot be applied to all cultures, as some starter organisms do not grow well in PIM. Furthermore continuous use of PIM can promote the development of insensitive phages (that is phages that can cause infection even when Ca^{2+} levels are very low).

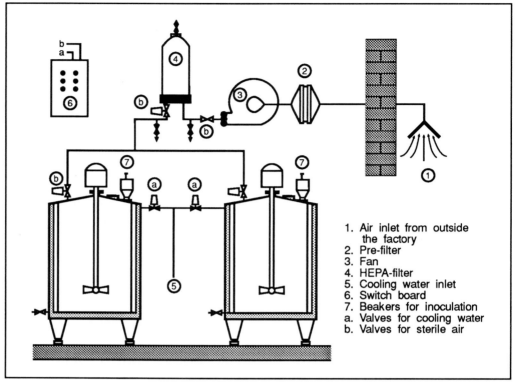

Figure 4.5 Cultivation tanks for bulk starter with over-pressure of sterile air and provided with inoculation beakers.

4.9.2 Prevention of bacteriophage development during cheese-making

The critical number of bacteriophages in cheese milk which inhibit acid production may vary for the different types of cheeses but is in the range of 10^3-10^4 ml^{-1} of cheese milk. Above such numbers, a retardation of the acid production in the young cheese will be observed. In modern cheese plants each vat will be used several times a day without rigorous decontamination between fillings.

∏   Do you think the reasons that vats are not decontaminated (by steaming) after each filling during a working day is based on scientific criteria or on economic criteria?

The answer is quite straight forward: decontamination is not done during the working day in order to save time (and hence money). Steaming and cooling takes up a great deal of time, and would greatly reduce output.

This means that by the end of the working day the level of bacteriophage contamination may have built up to above the critical threshold. To prevent this, the following procedures are presently common practice in many modern plants:

• the milk delivered to the dairy plant should have a low number of bacteriophages (less than one in 100 ml). In no circumstance should whey be transported in the same container or tank that is used to deliver the milk;

- if whey cream is used to standardise the cheese milk, it should be pasteurised (at least one minute at 90°C) to kill off any contaminating phages;

- inoculated milk should not be left in pipe-work between fillings, as phage can accumulate here. It is preferable to inject the starter and the cheese milk through separate pipework. Pipework should not have 'dead ends' in which milk or culture might remain trapped for some time;

- after each cheese-milk fermentation, the vat should be thoroughly rinsed out with bacteriophage-free (drinking quality) water.

SAQ 4.9

Consider how each of the factors listed below would affect the likelihood of phage infection leading to starter failure of a single-strain starter being used repeatedly as the sole starter in a cheese plant. If you think a factor would increase the likelihood of infection and failure mark '+' against the item. If you think the factor would decrease the likelihood of infection and failure mark the item with '−'. If you think the likelihood of infection and failure would be unchanged mark the factor with '±'.

1) High concentration of Ca^{2+} ions.

2) Repeated pre-exposure to phages contaminating the plant.

3) Addition of a phage-sensitive strain to the starter.

4) Insertion of pTR2030 plasmid into the starter.

5) The deletion of natural plasmids from the starter.

6) The addition of temperate phage to which the starter is sensitive.

SAQ 4.10

You are the Section Head of a cheese production unit. The unit has been operating for several years, using one single-strain starter, without failure due to phage infection. The Production Manager has been to a seminar on rotation of starters, is keen to try it, and has given you the options listed below for possible introduction in your production unit. Choose the option which would be best.

1) Carry on as at present with the currently used starter.

2) Rotate the currently used starter with a P-starter.

3) Rotate the currently used starter with a whey-derived starter.

4) Rotate the currently used starter with four other BIM starters. Thus the sequence would be:

Current strain ⟶ BIM_I ⟶ Current strain ⟶ BIM_{II} ⟶ Current strain etc

4.10 The prevention of development of bitter flavours

reduced rennet
results in
reduced bitters

We have already learnt that the combined action of the proteolytic activity from rennet and the starter bacteria may result in the production of bitter-tasting peptides from paracasein. These bitter peptides may, depending on the starter strain, be degraded into non-bitter compounds. Technological adjustments that result in a lower rennet content in cheese reduce bitter flavour development. These adjustments include: the addition of a minimum amount of rennet to cheese milk; the prevention of a low pH of the cheese milk at the moment of renneting; the application of high cooking temperatures. Other factors that affect the composition of the cheese and ripening may influence bitter flavour development. High salt content and low ripening temperature minimise development of bitter flavour in Gouda cheese. Other parameters such as pH and moisture content may also have an effect, but they are usually constant for each cheese variety.

The major factor controlling flavour development is the proteolytic system of the starter culture. The cell envelope-located proteinases of starter lactococci are known to have different specificities for casein and therefore different capacities to form bitter peptides. In addition, starter bacteria may contain peptidases that are able to degrade the generated bitter peptides. Based on these properties, a specific starter that does not produce bitter flavours may be selected.

high Prt⁻
starters

Another solution is to use a high percentage (eg 80%) of Prt⁻ strains in a mixed-strain or multiple-strain starter. This results in a relatively low concentration of proteinase that may produce bitter peptides in the cheese. Most of the strains used in the Dutch starter system are Prt⁻ strains. In addition, if in a single strain the capacity to produce proteinase is encoded by an unstable plasmid, it is very likely that such a strain will generate Prt⁻ mutants after several transfers. No more than 10% of strains in a culture need to be Prt⁺ in order to support growth at adequate rates and to allow for sufficient acid production.

temperature
sensitivity
influences
action of
proteinases
during
maturisation

In Cheddar and related cheese types, the development of starter cells is restricted by the cooking temperature (38-40°C) which is about 10°C above the optimum growth temperature (28-30°C). The temperature sensitivity of starters will thus influence the degree to which they will contribute to proteolysis. The pH and salt concentration also affect the proteinase activity.

4.11 The acceleration of cheese ripening

The present understanding of at least some aspects of cheese ripening has resulted in methods to shorten the ripening time (and thus reduce costs).

∏   See if you can make a list of the methods that could be used to accelerate the ripening of cheese. (Think about the processes which occurs during cheese ripening).

The main ways by which the ripening of cheese can be accelerated are:

•   the addition of proteases (and lipases) to the cheese. Addition would commonly be to the cheese milk;

- the addition to young cheese of proteolytic starters (or other organisms);

- the elevation of ripening temperature.

modified
starter cells

Although all these methods have been used and have specific advantages and disadvantages, this section will deal only with the addition of modified starter cells. The advantage of this is that no legal barriers are anticipated, which would prevent wide application of this method. In addition, the natural enzyme balance in the cheese is maintained and it is simple to incorporate the starter cells into the cheese.

At present the acceleration of cheese ripening by the inclusion of additional amounts of starter cells is still under development. However research in this field continues and expectations are high. Here the addition of foreign starter cells (those not normally used for the cheese variety in question) will not be discussed, since this may result in different cheese varieties rather than acceleration of ripening. Instead, attention will be focused on the addition of modified starter cells typical for the cheese variety.

heat shock
methods

The modification of additional starter cells is necessary, otherwise too much acid would be produced (which would have a detrimental effect on the cheese quality). A heat shock (short-term change in temperature) can prevent bacteria from producing acid while having less effect on proteolytic capacity. Heat-shock methods have successfully been employed in the accelerated ripening of Swedish and Dutch cheese. In the Dutch cheese trials, an additional 5% of a thermoshocked culture was added. Two to four times more cell mass was incorporated into the cheese curd than in the control made in the traditional way. A limited increase in the amount of liberated amino acids was observed, indicating extended proteolysis. The cheese had a less bitter taste and showed a moderately increased intensity of the typical Gouda cheese flavour. Ripening time was reduced by approximately 25%. The addition of more than 5% of a thermoschocked culture resulted in too extensive acid production during curd making. Applying heat shocks of different intensity to the starter culture led to the conclusion that a complete inactivation of the acid-producing capacity and a limited inactivation of the proteolytic system could not be achieved simultaneously.

∏ What type of genetic modification of added bacteria do you think would be necessary to overcome the problem of continued acid production during cheese ripening?

A Lac⁻ mutant would be unable to ferment lactose. It is feasible that a temperature-sensitive organism could be developed, which ferments lactose during fermentation of cheese milk, but could not ferment lactose at the different temperature of ripening. A possible mechanism for this would be a heat-labile or cold-labile lactose permease system.

Experiments in Australia in which the number of Lac⁻ cells incorporated into the cheese curd was 2-10 fold higher than in the control cheese, have shown that the rate of proteolysis in the cheese was increased. A good correlation was found between the production of free amino acids and the overall flavour intensity. However, the Lac⁻ mutants used so far imparted an atypical flavour to the cheese. These experiments were performed with *L.lactis* subsp. *lactis* strains. Since it is experienced frequently that *L.lactis* subsp. *cremoris* strains yield fewer atypical off flavours, experiments with Lac⁻ mutants of this subspecies are being undertaken to evaluate their potential for accelerated cheese ripening.

SAQ 4.11

In column A are listed factors which influence lactic acid production from starters. For each factor, list the stage(s) in the cheese making process listed in column B in which you think the factor exerts an effect.

A

a) Salt
b) Absence of lactose
c) High oxygen concentration
d) Concentration of starter
    organisms present

B

i)   Cheese-milk fermentation
ii)   Coagulation
iii) Cooking
iv)   Pressing
v)   Ripening

SAQ 4.12

In column A are listed properties of lactic acid bacteria. For each property, match it with the organism/starter listed in column B with which that property is associated, and select from column C the stage(s) in the cheese production process in which the property is important.

A	B	C
i) Lactic acid production at 39°C	a) Whey-derived starter	1) Cheese-milk fermentation
ii) Lack of proteinase activity	b) Aromabacteria	2) Coagulation
iii) Carbon dioxide	c) Prt⁻ mutant	3) Cooking
iv) Phage-insensitivity	d) Temperature-insensitive mutant	4) Pressing
		5) Ripening

SAQ 4.13

Listed below are some of the steps in the process to produce a frozen starter concentrate. List the steps in the order in which they would be carried out, starting with the first and finishing with the last.

Centrifugation.

Packaging.

Medium sterilisation.

Cooling to 1°C.

Incubation (20°C 20h)

Inoculation with stock culture.

Freezing to -35°C.

Incubation (20°C 20h).

SAQ 4.14

Answer true or false to each of the following.

1) Proteinase activity by heat-shocked organisms can lead to a reduction in formation of bitter flavours.

2) Proteinase activity is beneficial to the development of cheese flavour, as it degrades bitter tasting peptides formed by peptidases.

3) Proteinase activity may or may not be responsible for the production of bitter flavours: it depends upon the activity of the particular enzyme.

4) Proteinase activity can accelerate ripening of cheese by increasing the rate of formation of flavoured compounds (formed by degradation of protein).

Summary and objectives

In this chapter we have examined the use of starter cultures of lactic acid bacteria to make cheese. Different cheese types use different starter cultures with different metabolic activities. The ability of starter organisms to ferment lactose and degrade protein is particularly important. We have examined the factors which affect starter activity and the various procedures used to manipulate and optimise this. Various starter systems have been developed, mainly to minimise the risk of bacteriophage contamination. Selected starter cultures are produced in concentrated forms for supply to cheese factories. Now that you have completed this chapter you should be able to:

- draw and interpret a flow diagram describing the production of Gouda and Cheddar cheeses;

- describe the composition of the various types of starter cultures;

- describe some of the metabolic activities of micro-organisms important in cheese production;

- describe the factors affecting the activity of starter cultures;

- outline mechanisms of resistance to phage infection by starter cultures;

- describe operational procedures used to minimise the problem of phage infection in starter propagation and cheese-making;

- describe the various types of starter culture systems;

- describe methods for the genetic improvement of starter cultures;

- describe applications of genetically improved starter organisms;

- draw and interpret a flow diagram to describe the production of starter cultures;

- outline methods for the prevention of development of bitter flavours;

- outline methods for the acceleration of cheese ripening.

Chymosin: Production from genetically engineered micro-organisms.

Chymosin: Production from genetically engineered micro-organisms.

5.1 Introduction

The production of cheese was outlined in Chapter 3, and the role and production of starter cultures was described in Chapter 4. Cheese was probably first made thousands of years ago and was possibly the first process in which man applied enzymes and bacteria. Not only was a new type of food produced, but a valuable food source (milk) could be preserved by its conversion into cheese. In this chapter we are going to examine the cheese making process, and in particular the role played by rennet and rennet substitutes. We will examine in detail the development of a process to produce the rennet enzyme chymosin through the use of genetically engineered micro-organisms.

Rennets are relatively cheap enzymes and they represent the largest single industrial usage of enzymes, with a world market of about 25×10^6 l of rennet per annum, worth about US $100m. It is not surprising then that rennets are attractive to industrial enzymologists and biotechnologists.

5.2 Cheese manufacture

We have already discussed the manufacturing process for cheese in some detail (see Chapters 3 and 4). However we will review the process again, stressing the role played by rennet. Cheese making is a traditional way of preserving a perishable foodstuff by salting, and lowering the pH by lactic acid fermentation. It involves essentially a dehydration process in which the casein proteins, fat and colloidal salts of milk are concentrated 6-12 fold, with removal of 90% of the water of milk and essentially all of the milk sugar (lactose), whey proteins and soluble milk salts. Concentration is achieved by coagulating the casein milk proteins, either by acidification (acid cheeses) or, more commonly additionally by limited specific proteolysis (rennet cheeses). Acid cheeses are usually eaten freshly while rennet cheeses are usually ripened after manufacture develop the typical characteristics of the individual cheese types.

acid cheeses
rennet cheeses

All rennet cheeses are produced by an essentially common process. Cheese making begins with the addition to milk of a 'starter' culture of lactic acid bacteria. The starter converts the lactose to lactic acid, producing acidification *in situ* to about pH 5. In addition, milk-clotting enzymes known as rennet are added, which results in the formation of a coagulated casein gel, trapping the milk fat and milk proteins. The most important enzyme in traditional calf rennet is chymosin (derived from the Greek word 'chyme' meaning gastric liquid), also called rennin. This clotting usually takes about 20 minutes at pH 6.5 and 30°C for Gouda cheese. The gel is cut, resulting in a curd mass from which the liquid whey is removed by drainage and pressing. Expulsion of water is referred to as syneresis. The curd then undergoes further processing into the final cheese product. During subsequent ripening the typical flavour and texture of individual cheese types develops. Some types of cheese are made with alternative

chymosin

syneresis

rennets or with micro-organisms other than lactic acid bacteria, such as propionic acid bacteria, corynebacteria and certain types of mould.

5.2.1 The milk protein system

milk proteins

Milk is a complex biological fluid consisting mainly of water, fat, proteins, lactose, citric acid and inorganic components (primarily calcium phosphate). Milk proteins and their derivatives are the largest category of industrial proteins produced in The Netherlands annually; they are a popular ingredient in foods because of their high nutritional value and their favourable functional characteristics such as flavour and emulsifying action.

the milk proteins

Bovine milk, and probably the milks of all other species, contains two markedly different groups of proteins: the caseins and the whey proteins (Table 5.1). Caseins predominate and represent about 80% of total milk protein (approximately 27g l^{-1} of milk). The major casein proteins are $\alpha_{s1\text{-casein}}$, $\alpha_{s2\text{-casein}}$, β-casein and κ-casein, in the approximate ratio 4:1:4:1. It is the caseins which end up in the curd during the manufacturing of cheese. The remaining proteins are contained in the by-product whey. The predominant component is β-lactoglobulin. The other major whey proteins are β-lactalbumin, bovine serum albumin and immunoglobulins. The composition of milk proteins is shown in Table 5.1.

Protein	Approximate proportion of total protein (%)	Phosphate groups per molecule
Caseins:		
α_{s1}-casein	32	8
α_{s2}-casein	8	10-13
β-casein	32	5
κ-casein	8	1-2
	80	
Whey proteins:		
β-lactoglobulin	12	0
β-lactalbumin	4	0
immunoglobulins	3	0
serum albumin	1	0
	20	

Table 5.1 The important milk proteins. α - alpha, β - beta, κ - kappa.

Π Since cheese is composed mainly of the proteins and fats from milk, the quantity of cheese produced from a given quantity of milk (the cheese yield) could be increased if the whey proteins were included in the curd. Which method do you think would be most appropriate for this? (Choose from the following list - ammonium sulphate precipitation; ion exchange chromatography; electrophoresis; ultrafiltration; heat coagulation).

Ultrafiltration is the most appropriate method of separating whey proteins. It does not involve addition of any chemical to the whey, and can be operated in continuous processes. It is also the cheapest and can be operated on a large scale. A disadvantage of including whey proteins is that they would reduce the overall cheese flavour.

differences between caseins and whey proteins

Caseins and whey proteins differ in various ways. Caseins do not seem to have such a compact, folded structure as most proteins (including whey proteins), as a result of which they are more easily accessible to proteolytic enzymes. In addition, whey proteins denature (unfold) below 100oC whereas caseins are heat-stable at temperatures well in excess of 100oC. Furthermore, all the caseins are phosphorylated, (Table 5.1) whereas whey proteins are not. Because of their high phosphate contents the caseins bind calcium ions strongly. In milk, which contains about 0.1% calcium ions, the majority of the caseins exist as large particles known as casein micelles. These micelles consist, on a dry weight basis, of about 94% protein and 6% inorganic components collectively called colloidal calcium phosphate. The micelles are roughly spherical and highly hydrated. They have an average diameter of 100nm and contain about 5000 casein monomers. Together with fat globules these casein micelles scatter light considerably and produce the "milky" appearance of milk.

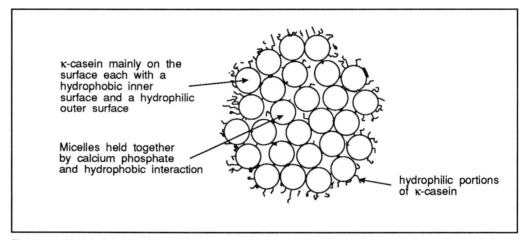

Figure 5.1 Model of the casein micelle.

internal composition of casein micelles

Although the structure of the micelles is not yet fully established, the general agreement is that they consist of spherical sub-micelles which are held together by colloidal calcium phosphate, hydrophobic interactions and hydrogen bonds. The α_{s-} and β-caseins are considered to form the hydrophobic core of the sub-micelles, while κ-casein is considered to be located predominantly on the surface of sub-micelles, forming a protective layer which stabilises micelles against flocculation. The N-terminal two-thirds of the κ-casein polypeptide chains are hydrophobic and presumably interact with the core proteins, leaving the hydrophilic C-terminal tail of about 70 amino acid residues projecting into the aqueous surrounding environment. These hydrophilic protuberances would give the micelles a 'hairy' appearance, promoting solubility and stabilising the micelle by preventing close approach of the sub-micelles.

5.2.2 Primary phase of coagulation (rennet action)

mode of chymosin action

The stability of micelles is promoted by κ-casein, which stabilises the α_{s-} and β-caseins. Without this stabilisation, the α_{s-} and β-caseins are precipitated by the action of calcium ions. Acid proteinases in rennet cause the rapid and specific hydrolysis of κ-casein. Chymosin, the major acid protease, acts mainly on the bond between the Phe-Met amino acid residues at positions 105 and 106 on the κ-casein molecule (Figure 5.2) producing para-κ-casein and glycomacropeptide fragments.

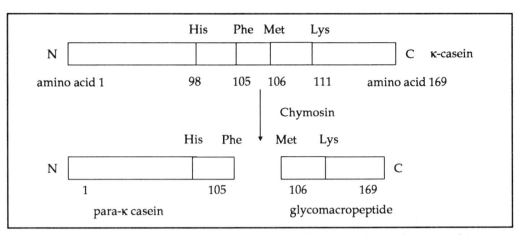

Figure 5.2 Diagrammatic representation of the action of chymosin on κ-casein (not drawn to scale).

The specificity of chymosin action is determined by the structure of the enzyme. This enables it to recognise the sequence of amino acids His 98 - Lys 111. This sequence has the correct conformation to fit into the active site cleft of acid protease enzyme molecules, thus the specificity of chymosin is such that it reacts specifically with this sequence in κ-casein.

5.2.3 Secondary phase of coagulation (flocculation)

Hydrolysis of κ-casein causes the release of its highly charged, hydrophilic C-terminal peptide, the glycomacropeptide. We have already seen that this component is responsible for the stabilisation and solubility of micelles (section 5.2.1). Removal of the glycomacropepides removes the protruding 'hairy' peptides from the sub-micelle surface and so reduces the repulsive forces which stabilise the micelle (these forces are both electrostatic and steric). When 85% of the κ-casein has been hydrolysed the micelles begin to participate in gelation, and coagulation of the milk will become apparent.

gelation

The coagulation of milk by the action of rennet is dependent upon several factors. The rate of coagulation generally increases with increasing temperature and, for bovine milk, will not take place below 15oC. Calcium ions are essential for coagulation, and so calcium chloride is added to milk (approximately 10g 100 l^{-1}) to reduce the clotting time. It also produces a stronger gel (or curd) which helps to improve the cheese yield. The concentration of salt and the pH also affect coagulation rates.

Π Choose the correct completion to the following statement.

The effect of adding a bivalent ion complexing agent (such as EDTA or citrate) to milk during coagulation by rennet would be to:

1)   increase the cheese yield;

2)   not affect the cheese yield;

3)   decrease the cheese yield;

4)   inhibit coagulation.

The correct answers are 3) and 4). Calcium ions are essential for casein micelle integrity and hence for coagulation. If calcium ions are removed by a complexing agent you would expect the casein micelles to dissociate into sub-micelles which will not coagulate.

∏   Excessive heating of milk prior to cheese making leads to denaturation of whey proteins, which then bind to micelles, and in particular to κ-casein. What effect do you think this might have on cheese yield?

If denatured proteins bind to κ-casein micelles this can have the effect of increasing the amount of coagulated protein in the curd. This leads to an increased cheese yield but of poor quality.

measure of rennet activity   It is essential that we have a standardised method for measuring and comparing rennet activities. Since we use rennet to coagulate milk, this property is usually used to measure rennet activity. One unit of activity clots 1 ml of milk substrate in 100 seconds under standard conditions of pH, temperature and calcium concentration.

5.2.4 Proteolysis during cheese ripening

development of flavour   Acid-coagulated cheeses are usually consumed fresh but the vast majority of rennet-coagulated cheeses are ripened (matured) for periods ranging from 4 weeks to 2 years. The duration of ripening is more or less inversely proportional to the moisture content of the cheese. During ripening, a multitude of microbiological, chemical and biochemical events occur, as a result of which the milk caseins, lipids and residual lactose (the main constituents of the cheese) are degraded to primary and later to secondary products. Among the principal compounds that have been isolated from several cheese varieties are: peptides, amino acids, amines, acids, thiols and thioesters (from proteins); fatty acids, methyl ketones, lactones and esters (from lipids); organic acids (acetic, propionic, lactic acid), carbon dioxide, esters and alcohols (from lactose). In certain combinations, these compounds are responsible for the characteristic flavour of various cheeses.

enzymes responsible for proteolysis   While lipolysis (fat degradation) is critical only in certain cheese varieties, glycolysis (sugar degradation) and proteolysis (protein degradation) are essential in all varieties. Proteolysis in cheese is a highly complex process due to the numerous types of proteolytic enzymes involved. These are primarily the milk-clotting enzymes in rennet (essentially chymosin and pepsin) and the proteolytic enzymes of the lactic acid bacteria, but in addition the native milk proteinases (essentially plasmin). Enzymes of added secondary starters such as propionic acid bacteria, yeasts and moulds may also contribute to milk protein degradation.

The initial products of casein proteolysis are oligopeptides, dipeptides and amino acids. Since these are non-volatile, they do not contribute to the odour or aroma of dairy products, only to the flavour. However, further reaction products generated by other enzymes are volatile, affecting both the flavour and aroma of cheese.

Depending on the pH, temperature and type of rennet used, between 5-20% of the rennet enzymes added to cheese milk are retained in the curd, the rest are lost in the whey. Nevertheless the amount of retained milk-clotting enzymes and their residual activity can be crucial in determining the quality of the cheese produced. This is important to remember when considering the use of rennet substitutes (see later).

∏ The use of rennet enzymes is not very efficient since they are only used once and then lost in the whey. Before reading on draw an outline scheme of how you think the enzymes of rennet could be used more effectively.

One way would be to recover the enzymes from whey and re-use them. Whey itself is a valuable by-product, and so recovery procedures should not alter its composition. Another way would be to immobilise the rennet enzymes on a solid support and pass the milk through the bed of immobilised enzyme continuously. The key question here would be, would the immobilised rennet enzyme act efficiently on the casein micelles as the milk is passed through?

5.3 Rennet and rennet substitutes

5.3.1 Biosynthesis of chymosin

Bovine chymosin (rennin, EC 3.4.23.4.) is an acid protease which is synthesised in the fourth stomach (abomasum) of young calves in the form of the precursor pre-pro-chymosin (Figure 5.3). Through a series of proteolytic cleavages of its amino-terminus, pre-pro-chymosin is first converted, during secretion into the stomach, to pro-chymosin comprising 365 amino acids and molecular weight (MW) of 40 500. Pro-chymosin is enzymatically inactive, but it is autocatalytically converted to an active enzyme at acidic pH. At pH 4.5, 42 N-terminal amino acids are removed, producing active chymosin of MW 35 600, while at pH 2.0 only 27 amino acids are removed, producing pseudo-chymosin of MW 37 400. Once formed, pseudo-chymosin is relatively stable at a pH below 3 or above 6, but it is further converted to chymosin by incubation at pH 4.5. Both pseudo-chymosin and chymosin are enzymatically active as measured by milk-clotting activity. Acid-catalysed conversion of pro-chymosin is referred to as activation; it does not necessarily require the activities of other enzymes, but it is accelerated by active chymosin.

Chymosin exists in the form of two isoenzymes (A and B) differing only at residue 244 (aspartic acid in the A form and glycine in the B form). Both allelic variants are found in calves throughout the world, but each individual calf will only have one form, either chymosin A or chymosin B. The overall distribution of chymosin A and B in calf rennets is about 40% A and 60% B and reflects the natural proportion of the A and B genes in calves. Interestingly, chymosin A has a 15-20% higher specific activity for κ-casein than chymosin B, but it is less stable.

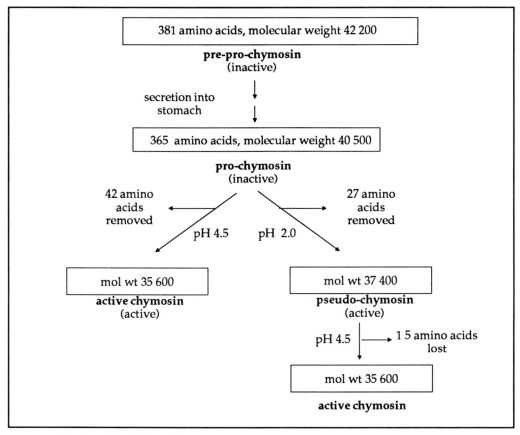

Figure 5.3 The formation of chymosin.

5.3.2 Bovine rennet

rennet of choice

Many proteinases will coagulate milk under suitable conditions, but most have a high non-specific proteolytic activity relative to their specific milk-clotting activity. Consequently, they hydrolyse the coagulum too quickly causing reduced cheese yields and/or defective cheese with a tendency to bitterness. Although plant proteinases (such as ficin, papain, bromelain) appear to have been used as rennets since prehistoric times, gastric proteinases from calves, young goats or lambs have been used traditionally as rennets. Calf rennet has been the rennet of choice in industrial cheese manufacture. It is prepared by soaking the calves stomachs in water. Salt is added to reduce the water activity and so suppress bacterial contamination. The liquid infusion is later drawn off. It is this which constitutes rennet.

alternatives to calf rennet

Rennet from cow's stomach contains chymosin and pepsin. The proportion of chymosin is highest (at least 90% of clotting activity) in the youngest calves and decreases with age to less than 10%. Owing to increasing world cheese production (about 4% increase annually over the past 20 years) and a reduced supply of calf rennet (due to a decrease in calf numbers and a tendency to slaughter more mature calves), the supply of this chymosin-rich calf rennet has been inadequate for many years. This has led to an increased price for calf rennet and to a search for rennet substitutes.

5.3.3 Rennet substitutes

Despite the availability of numerous potentially useful milk coagulants, only six rennet substitutes have been found to be more or less acceptable for one or more cheese varieties. These are the pepsins from cow, pig and chicken and the fungal acid proteinases from *Mucor miehei* (some trade names: Fromase, Hannilase, Marzyme, Miki. Rennilase), *Mucor pusillus* (trade names: Emporase, Meito, Noury, Novadel) and *Endothia parasitica* (trade names: Supraren, Surecurd).

animal substitutes

Chicken pepsin is the least suitable of these and used widely only in Israel and to a lesser extent in Czechoslovakia. Bovine pepsin is probably the most satisfactory and many commercial calf rennets contain about 50% by weight of bovine pepsin. The initial proteolytic specificity of this enzyme is similar to that of calf chymosin and it gives generally satisfactory results with various cheese types. However, bovine pepsin has a much higher level of non-specific proteolytic activity which weakens the protein network needed to entrap milk fat during coagulation and leads to decreased cheese yield. The activity of porcine pepsin is very sensitive to pH and it may be extensively denatured during cheese making and consequently contribute little to proteolysis during cheese ripening.

fungal rennets

microbial rennets

The overall proteolytic specificities of the three commonly used fungal rennets are considerably different from that of calf rennet, but the acceptability (palatability) of most young cheese varieties made using fungal rennets has been fairly good. Microbial rennets are widely used in the United States and France but calf rennet is still predominant in most other European countries and New Zealand.

∏ Although fungal rennets and high-pepsin rennets are suitable for the production of young cheese (matured for a short time) they are less suitable for the production of matured cheeses. Why do you think this is so? (Refer to section 5.2.4 for a clue).

The fungal and high-pepsin rennets have different (and generally greater) non-specific proteolytic activities compared to rennet. This activity during ripening can lead to the production of flavours and odours which are undesirable. In young cheeses the maturation period is short, so the undesirable changes do not have time to occur.

5.3.4 Specificity of rennet enzymes

Chymosin and all the commercially successful rennet substitutes are acid proteinases. Their amino acid sequences and their 3-dimensional structures are highly similar and they can be regarded as a family of similar enzymes. Acid proteinases have relatively narrow specificities with a preference for peptide bonds to which bulky hydrophobic residues supply the carboxyl group. This narrow specificity is significant for the success of these enzymes in cheese manufacture. Calf chymosin, porcine pepsin and *M.miehei* proteinases have all been shown to preferentially cleave and Phe 105-Met 106 bond of κ-casein, while the proteinase of *E.parasitica* cleaves the adjacent Ser 104-Phe 105 bond. The fact that, in cheese, these enzymes operate at pH 5.5 to 6.8 - far removed from their optima (about pH 2.0 for pepsins) - is probably also significant. Not all acid proteinases are suitable as rennets, because many are too active even under the prevailing relatively unfavourable conditions.

SAQ 5.1	Answer true or false to each of the following statements. 1) Rennet is added to milk because acidification (by lactic acid bacteria) alone cannot produce coagulation. 2) Rennet enzymes help to produce a characteristic type of cheese by participating in ripening. 3) Rennet is added to milk as the milk-clotting enzymes from lactic acid bacteria cannot coagulate milk on their own.

SAQ 5.2	Answer true or false to each of the completions to the following statement. For the manufacture of old (long ripening) cheeses, rennet from young calves is preferred to bovine rennet because calf rennet: 1) contains a higher proportion of chymosin: pepsin; 2) produces higher non-specific proteolysis during ripening; 3) produces firmer gels during coagulation; 4) produces greater degradation of κ-casein during coagulation.

SAQ 5.3	Which of the following could produce the fragment 105 Phe-Met-Ala-Ile-Pro-Pro-Lys...169 from κ-casein? 1) Calf rennet. 2) A mixture of bovine chymosin A and B. 3) Protease from *Mucor miehei*. 4) Protease from *Endothia parasitica*.

SAQ 5.4	Which of the following enzymes are most important for coagulation and which for ripening of rennet cheeses? 1) Rennet enzymes. 2) Native milk proteinases (plasmin). 3) Proteolytic enzymes of lactic acid bacteria. 4) Glycolytic enzymes. 5) Lipolytic enzymes.

5.4 Chymosin production from genetically engineered micro-organisms: A case study

We have already seen that rennets are economically valuable products, and that the supply from animal sources is limited. It was natural that, as techniques of genetic engineering were developed, attempts would be made to insert the gene for bovine chymosin into micro-organisms so that the enzyme could be made by fermentation. This possibility was made even more attractive by the fact that several enzymes (including some used in foods) were already produced from microbial fermentations, and so expertise on the fermentation and recovery of food-grade enzymes already existed.

We are going to examine how such a process was developed, from its inception to production. The general approach is common to any product, and you will be able to see this when you examine the other case studies in this book. Various stages in the process which are of importance are outlined below:

- planning stage - where questions are raised such as: can we make it; can we sell it; how can we make it; how much can we sell it for; how much can we sell; what will it cost us to make it; where will we make it;

- research stage - where laboratory-scale work is undertaken to develop a means of producing the product;

- development stage - where the production process is developed and optimised. It involves laboratory-scale and larger pilot-scale work. Research and development (R & D) are closely linked;

- product formulation and regulatory requirements;

- scale-up - where the process is transferred from pilot-scale to full-scale (production-scale);

- commercial operation - where the process is fully developed and operating in final form.

patenting

formulation
specification

quality control

Other vital operations fit into this overall scheme. Processes might need to be patented, in which case patents need to be drawn up and appropriately lodged. Processes and products might be covered by various legislation, which needs to be conformed to. Product formulation and specification are decided upon (that is what form the product will take (such as solid/liquid) and what quality control specifications it must meet. The product must also be marketed and an efficient distribution and sales network ensured. These operations are carried out in parallel with those listed above, usually around the time of scale- up. Let us now examine these aspects in relation to chymosin production.

5.4.1 The planning stage.

Due to increasing cheese production and decreasing slaughtering of suckling calves, a growing shortage of calf rennet developed. Alternative rennets produced by fungi and rennets with high pepsin content were introduced in the market, but as explained in the previous section, these alternatives are less suited for the production of long-ripened cheese. The demands of the market for rennet substitute were determined and can be summarised as:

- a standardised product of high quality;

- constant and guaranteed supply at a stable price;

- product to be cheaper than calf rennet.

By 1980 the development of recombinant DNA technology was at a stage that made it possible to pass the species barrier and transfer genetic information from one species to another. For example, it became possible to isolate bovine DNA that coded for the milk clotting enzyme chymosin, and transfer it to a micro-organism. The manipulated micro-organism could then be used for the production of the desired enzyme. Instead of extracting the enzyme from calf stomachs, it became possible to produce chymosin in a process based on fermentation and recovery technology. This method of production could meet the market requirements given above since:

- well-defined raw materials can be used, and so the composition of the rennet produced by fermentation is constant;

- the supply of chymosin produced by fermentation is not affected by variations in availability of calf stomachs;

- it was estimated that the production price of the chymosin produced in a fermentation process could be lower than the price for the extract of stomachs.

Based on these considerations a number of enzyme manufacturers decided to develop a process for the large-scale production of chymosin by means of a recombinant micro-organism. The specific approach for such a development depended on various factors which were different for different companies, and therefore a variety of development strategies were chosen. The evaluation of a project concept can be carried out from different points of view, for example economic, technological and strategic.

Economic: Will the product give sufficient return on investments?

high-risk

high-return

Before the product can be introduced into the market, money has to be invested for activities such as research and development, registration, patent applications and patent defence, adaptation of production equipment and marketing. These investments have to be earned back by the profits gained from the sales of the product. The profits will depend on the estimated market share that can be obtained and the profit margin on the product. In making the balance between investments and returns, you have to take into consideration that the project can fail before the break-even point has been reached. Therefore the estimated return on investments should be high for high-risk investments.

Technical: Are the objectives feasible?

competitors

First of all you have to predict whether the project concept is technologically feasible. In doing this you must consider various limiting factors like the costs of the development, the duration of the project and the criteria for the product quality imposed by the market. In addition to estimating the chance of success of the project itself, it is essential to judge its strength in comparison to options that might be chosen by competitors.

Strategic: Does the project fit in the company?

Based on an analysis of strengths and weaknesses, opportunities and threats, a company will decide on a strategy for the future. The implementation of the chosen strategy will call for certain specific developments. It will be clear that in reaching a decision to start a new development, it will be important whether or not the proposed concept fits the organisation's strategic plan. For example, the chance of success of the project will be higher in the case where it is based on the technological expertise of the company and has synergistic relations to other developments. Furthermore, one has to make sure that the relevant (technological, application and marketing) know-how is, or can be, made available.

The overall objective of a project for the development of a rennet product based on chymosin produced by a recombinant micro-organism will be more or less the same for different companies and for the purpose of this case study can be defined as:

- development (within a certain time) of a high-quality rennet product, based on production of chymosin with a recombinant micro-organism, that can be supplied in sufficient quantities at a competitive price.

5.4.2 Research stage

Development of the production strain

One of the first choices to be made in the development of a production process for chymosin by means of recombinant DNA technology, is the selection of a suitable host organism. Together with an expression cassette active in the chosen organism, another important issue is the choice of a gene vector. This obviously needs to be expressed in the chosen host as well as being able to carry the chymosin gene. It also should provide a mechanism for distinguishing between host cells and host cells containing the vector. Appropriate choices here would allow the development of a strain able to produce chymosin. After the relevant characteristics of this strain have been checked, it could be used for the large-scale production of the enzyme.

Selection of the host organism

E. coli

inclusion bodies

By far the most frequently used organism in genetic engineering is *Escherichia coli*. Therefore, from a technological point of view this would be the first organism to evaluate. It was found that, in this host, chymosin was produced in an inactive, insoluble form in so-called inclusion bodies. To isolate and activate the chymosin the use of protein denaturing agents would have been necessary. Another drawback of the use of *E. coli* as host organism was the fact that it is not commonly accepted for the production of compounds that are used for human consumption.

∏ Which of the following completions to the statement is/are correct? *E. coli* is not generally accepted for the production of food compounds because:

1) some strains of *E. coli* cause food poisoning;

2) bacterial products are not permissible in foods;

3) *E. coli* can produce endotoxins which are toxic to humans;

4) *E. coli* is not traditionally associated with food products.

1), 3) and 4) are correct answers. *E. coli* could potentially produce substances during fermentation which are toxic to humans. If these contaminated the product (or viable organisms remaining in the product, illness may result. 2) is incorrect. There are many examples of bacterial products found in food (eg lactic acid in cheese, acetic acid in pickles etc). If you chose 2) as a correct answer re-read Chapters 2 and 3.

Because of the difficulties encountered with *E. coli*, several micro-organisms that are already used for the production of food enzymes were tried as host organism. Some examples are given below.

Bacillus *Bacillus* strains are non-pathogenic and are used in the large-scale production of industrial enzymes like α-amylase. It was found that *B. licheniformis* could produce bovine pro-chymosin, but the signal sequences used did not allow efficient secretion of the enzyme into the culture medium.

Lactococcus Since lactic acid bacteria are used in production of cheese, it seemed obvious to choose them as host organism for the production of chymosin. Production of this enzyme has been achieved in *Lactococcus lactis*, but the production level was low.

Saccharomyces Intracellular production of chymosin in *Saccharomyces cerevisiae* (yeast) resulted in an insoluble product as in *E. coli*. By fusing a secretion signal-coding sequence to pro-chymosin, the production of activatable extracellular pro-chymosin in *S. cerevisiae* was demonstrated. However, the secretion efficiency was very low.

Kluyveromyces *Kluyveromyces lactis*, a yeast associated with milk products, has been used for many years for the production of the dairy enzyme lactase (β-galactosidase, see Section 2.4.2). Therefore, its characteristics and its performance in large-scale production were already well understood. It was found that, in contrast to *S. cerevisiae, K.lactis* was capable of

glycosylation synthesis and secretion of fully active pro-chymosin. Various signal sequences could be used to efficiently direct the secretion of pro-chymosin.

Using certain food-grade fungi, like *Aspergillus*, as host organism, could give the advantage of very high production levels. However, in certain cases glycosylation (addition of carbohydrate to the protein) may occur. In such cases the product would not be identical to the chymosin in calf stomachs, and in the development of a production strain additional work would be necessary.

The choice of the host organism turned out to be different for the various companies involved in the development of commercial recombinant chymosin. It will be obvious that this will have an important effect on the approach of the project. Some examples of host organism for recombinant DNA chymosin products are:

Escherichia coli	Pfizer, USA
Kluyveromyces lactis	Gist-Brocades, Holland
Aspergillus niger	Genencor, USA

Table 5.2 Organisms used for recombinant chymosin by various companies.

We will now examine the development of recombinant chymosin using *K. lactis* as host organism.

The requirements of DNA constructs for introducing selected genes into host organisms are well-understood and are covered elsewhere in this BIOTOL series. To refresh your memory the process is summarised in Figure 5.4.

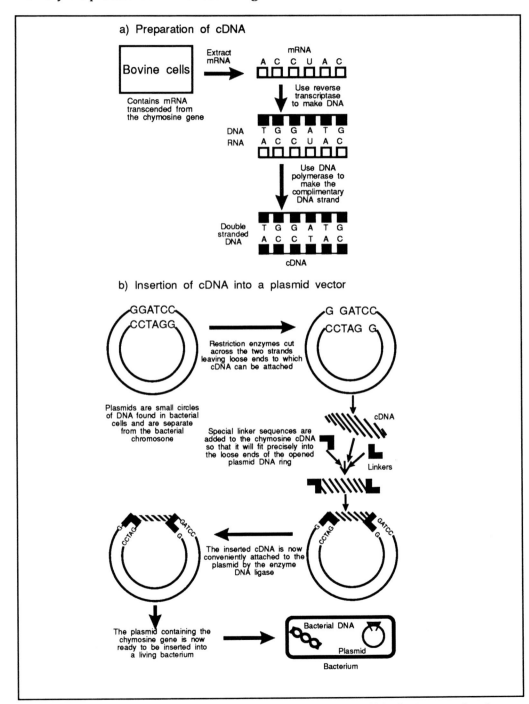

Figure 5.4 The overall process of chymosin production using recombinant *K. lactis*. a) preparation of cDNA, b) insertion of cDNA into a vector.

reverse
transcriptase

expression
cassette

The core of the DNA construct that will be introduced in the selected host organism is obviously the part that contains the information for the chymosin molecule. This can be obtained by isolation and enrichment of the messenger RNA (mRNA) of the desired product followed by conversion into double stranded DNA. This can be achieved using the enzyme reverse transcriptase and DNA polymerase. In order to secrete the enzyme into the fermentation medium, a DNA- sequence coding for a leader sequence is put in front of the chymosin gene. Together with suitable promoter and terminator sequences, this will make a 'cassette' for the expression of chymosin.

Such a cassette can be incorporated into a plasmid of the host organism. Plasmid-based expression systems can be very unstable. It would be better if the required gene was integrated into the chromosome of the hosts. Such a system would be more stable and would give an important advantage. For the integration in the *K. lactis* chromosome, the locus of the lactase gene was chosen.

In order to be able to select the individual *K. lactis* cells in which the construct has been integrated, a selection marker was incorporated. A typical example of one of these vectors is shown in Figure 5.5.

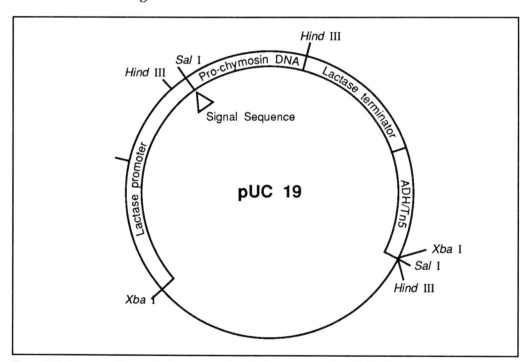

Figure 5.5 An expression vector for incorporation of chymosin gene into *K. lactis*.

For the different functional parts of the expression vector, there are several options for the actual construction. Let us consider four of these.

choice of the
gene

• Which gene should be used - that for chymosin or pro-chymosin? The choice that has to be made is whether the sequence for the active enzyme or for pro-chymosin is to be used.

∏ Can you think of any advantages which would result from producing pro-chymosin as a product compared to chymosin?

Chymosin promote proteolysis, which might (during fermentation, recovery and storage) cause disturbances to the producing organism and autocatalytic breakdown of the enzyme itself. Pro-chymosin is inactive, and so does not have these disadvantages.

Another aspect concerning the choice of the gene is the heterogeneity of chymosin. Remember there are two forms (A and B) found in calves. In cheese making little difference between the two forms is experienced. As a matter of fact commercial products have been developed based on both A and B forms.

• Which protein leader sequence gives the best secretion?

choice of leader sequence

Several protein leader sequences have been tested to direct the secretion of pro-chymosin by *K. lactis*. First of all the homologous signal sequence of the chymosin was tried. The pro-chymosin was efficiently secreted into the culture medium. Secondly, leader sequences from yeast proteins were used, in particular the pre-pro-region of the α-factor genes from *S. cerevisiae* and *K. lactis*. More than 95% of the pro-chymosin produced was found in the culture medium. Also other leader sequences from yeast proteins were successfully used in *K. lactis* to secrete the desired enzyme.

• Which promoter and terminator are to be chosen?

choice of promoter and terminator

Again there are numerous options. Since *K. lactis* is used on a commercial scale for the production of lactase, the regulation of lactase production in this organism was quite well understood. Therefore, the pro-chymosin gene plus a leader sequence were put under the control of the lactase promoter and terminator. It turned out that the regulation of the production of the heterologous enzyme was comparable to the regulation of lactase.

• How should the expression cassette be integrated into the yeast chromosome?

integration into host chromosome

By using the *K. lactis* lactase promoter in the construct, a common sequence was available in both the construct and the chromosome of the host organism. The expression vector was targeted to the lactase gene of the host cell by linearising the construct at a unique site in the lactase promoter. The vector is integrated into the host chromosome by recombination during DNA replication.

Characteristics of the production strain

Whether or not a specific organism is suitable for the production of enzymes on a commercial scale will depend on various factors. These include the medium requirements, fermentation process conditions and the behaviour in down-stream processing of the fermented medium.

changed characteristics of recombinant strains

The characteristics of the production strain are of course partly comparable to the characteristics of the host strain. However, in the early years of the development of recombinant DNA organisms for industrial processes, little was known about the effect of the introduction of heterologous DNA on the general characteristics of the organism. Therefore, extensive tests had to be done to check all conceivable alterations. As we will see later on, for the chymosin-producing *K. lactis* strain no changes other than those introduced by the DNA construct were experienced.

stability of
chymosin
production

An important factor in the evaluation of the strain was the stability of chymosin production. This was tested by growing the strain in seven consecutive shake-flask cultures, and determining the chymosin content in these cultures. It was shown that no significant decrease of production capacity occurred during at least 45 generations, which is sufficient for large scale production. With Southern Blotting (a technique for determining changes in nucleotide sequences) it was confirmed that no rearrangements had occurred in the integrated expression cassette and that its copy number was stable.

Another method to check the stability of the chymosin producing capacity of an industrial strain would be to isolate a single cell from the broth at the end of the fermentation and use it as an inoculum for a new fermentation. The amount of product formed should be equal in both cases.

Biochemical identity of the enzyme

SDS-gel
electrophoresis

Once a recombinant microbial strain capable of producing chymosin had been developed, it had to be proven that the product was identical to the chymosin found in the extract of calf stomachs. A rapid method to detect a possible alteration in the molecular structure of the enzyme was by SDS-polyacrylamide gel electrophoresis. This method can also give some indication of the presence of possible degradation products.

HPLC gel
filtration
immune assay

The molecular size of the native form of the product can be checked by HPLC gel column filtration, against the chymosin purified from calf rennet. Furthermore, using monoclonal antibodies targeted against calf chymosin, the immunological characteristics can be determined.

⫠ Figure 5.6 shows the result of comparisons of various rennet preparations by SDS-gel analysis. What can you conclude from this about the recombinant chymosin (MaxirenR)?

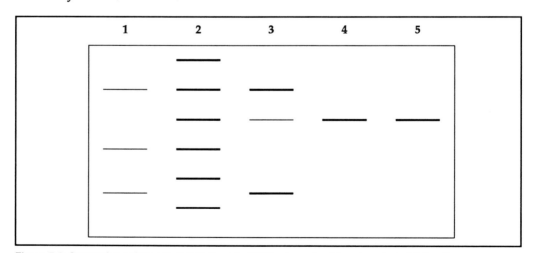

Figure 5.6 Comparison of rennet preparations by SDS-polyacrylamide gel analysis (stylised).
Lane 1: commercial USA rennet, lane 2: molecular weight markers, lane 3: commercial Dutch rennet, lane 4: chymosin purified from commercial Dutch rennet, Lane 5: Maxiren[R] recombinant chymosin.

The conclusion is that the recombinant chymosin has the same molecular weight as the native bovine chymosin in rennet.

∏ From the SDS-polyacrylamide gel analysis shown in Figure 5.6, does the commercial USA rennet contain an enzyme of comparable molecular weight to native bovine chymosin?

You should have come to the conclusion that commerical USA rennet does not contain an enzyme of comparable molecular weight as native bovine chymosine.

In addition to the above mentioned biochemical analysis of the intact molecule, a number of specific points had to be checked in more detail.

From the DNA sequence introduced into the yeast chromosome it was expected that a protein with an amino acid composition and sequence identical to one of the allelic forms of chymosin would be produced.

∏ Write down analytical methods you would use to compare 1) the amino acid composition and 2) the amino acid sequence of recombinant chymosin with that of native chymosin. When you have completed this compare this with our suggestions.

1)   The amino acid composition of the protein would involve a) complete hydrolysis to constituent amino acids b) analysis of the quantity of each amino acid by appropriate chromatographic techniques (such as HPLC).

2)   Sequencing of the amino acids involves the hydrolysis of the protein using specific enzymes to create specific fragments. The analysis of the amino acid sequence in each of these fragments can be used to build up a picture of the amino acid sequence of the complete protein.

check for glycosylation

Another potential problem in producing chymosin in fungal systems is glycosylation of the protein. Chymosin is not naturally glycosylated, so the recombinant product had to be checked for the presence of unwanted saccharides. Chymosin from calf rennet and fermentation broth from recombinant *K. lactis* were extensively purified, and assayed for sugar residues (using a sensitive method based on thiobarbituric acid). The results indicated that neither product was glycosylated.

checks for functional properties

From biochemical investigations such as those described above, the identity of the chymosin in calf rennet and that produced by a recombinant micro-organism were established. In addition to this, the functional properties were checked. For practical reasons, this was usually done by comparing the final product with animal rennet in relevant application tests. This will be discussed in a later section.

5.4.3 Development stage

In the previous section concerning the development of a recombinant DNA organism capable of producing chymosin, the actual production process was not specified. In the first phase of the genetic engineering process this was indeed not very important, since the main goal was to simply obtain a strain that could produce the desired product. Shake flask experiments were sufficient to prove that this goal has been reached. The next step was the development of a production process. In practice, the strain and the production process were developed as parallel activities.

The strategy concerning the design of the production process may differ significantly between different companies. The choice of how to develop the process will be influenced by the host organism chosen.

The choice of *K. lactis* as a host organism was, apart from the compliance to criteria for food additives, based on the availability of large scale production know-how with this organism for the production of lactase. As discussed earlier, in the design of the expression cassette, the lactase promoter and terminator sequences were also chosen. The existing lactase production protocol was used as a starting point for the process development. Although a number of adaptations were introduced to optimise the process, the general process remained the same (Figure 5.4) as for lactase production. The concept of using well-known host organism in which special genes are easily 'plugged' in has been named the PLUG- BUGTM concept.

Let us consider four important aspects of the development process.

The fermentation medium

An industrial medium containing all compounds needed for the growth of the strain and enzyme production, was available from the lactase production process. Since the composition of the medium influences further steps in production, some alternatives were tried in relation to the product recovery and formulation. Since chymosin is secreted into the medium during the fermentation process, the stability of the enzyme in the broth had to be checked.

The fermentation process

process optimisation

fed-batch specific productivity

The optimisation of the fermentation process was not fundamentally different from a process with a non-recombinant micro-organism. Starting from the protocol for the lactase process, the fermentation parameters were refined. In particular, pH, temperature and aeration were finely tuned to optimise yield. The fermentation was usually a fed-batch process. The rate of addition of the limiting substrate, often the carbon source, determined the growth rate of the organism. Through controlling the rate of addition of the feed, the optimal profile for the production of the desired enzyme was found. There were however some notable differences between lactase and chymosin production. In the production of the intracellular enzyme lactase, the productivity per weight of biomass (the 'specific productivity') decreases at the end of the fermentation. In the extracellular production of the recombinant chymosin, this is not the case.

yields vs process time

Other factors which influence process design include the size of the available plant capacity. Where the available plant capacity is adequate, the most important target will be minimum production cost per unit of product (in other words to make the product as cheaply as possible). This is achieved by reaching a compromise between high yields on raw materials and a short fermentation time: when the fermentation period is short, medium ingredients remain unused, whereas when the fermentation period is longer (with higher running costs) the medium ingredients will be more fully utilised. Where the plant capacity is limiting, the 'output' (volume of product produced per unit time) can become an important target. Output can be maximised (for batch or fed-batch cultures) by shortening the fermentation period. This might be necessary, even if the cost per unit product is increased due to wastage of medium ingredients. Medium ingredients contribute significantly to production costs.

Inactivation of the production organism

The production organism should be totally absent in the final product. In the case of *K. lactis* it was decided to kill the yeast before the fermenter was opened. By introducing this killing step, the downstream processing could then be essentially the same as that of any enzyme produced by fermentation.

∏ Why do you think the recombinant organism needs to be killed before the contents of the fermenter are processed?

Choose from the following list 1) for legal reasons, 2) to prevent rivals stealing the strain, 3) to prevent the culture from contaminating other cultures being used in the environs of the factory.

The main reason relates to legal requirements.

Regulations limit the degree to which genetically engineered organisms can be released into the environment. It would be undesirable to contaminate the equipment, the operators and the factory environment with such viable organisms. Killing the organisms while they are contained within the fermenter overcomes this problem.

n-octanol as a killing agent
In the initial process design the organisms were killed by addition of n-octanol, which causes to the lysis of the yeast. In this way all intra- and extra-cellular chymosin was made available in the liquid phase of the broth. The n-octanol had to be removed in a subsequent recovery process (which meant an additional step). Furthermore the lysis of the yeast also introduced cell components into the medium.

benzoate as a killing agent
For strains that secreted almost all the chymosin into the medium during the fermentation, complete autolysis of the cells is not necessary. It was found that in this case, the yeast cells could be easily killed by addition of benzoic acid at a low pH value. Benzoic acid is in any case a commonly used preservative for rennet products. Therefore, a saving in the final formulation step was obtained. The low pH of the mixture used during the killing of the yeast also facilitated the autocatalytic conversion of pro-chymosin into chymosin, similar to the natural processing occurring in the calf stomach.

Product isolation

filtration
The yeast cells are killed but not lysed. The morphologically intact cells can be removed easily by an appropriate filtration method, such as by rotary vacuum filtration. Filtration aids may facilitate this process. The filtrate and the wash water are combined and contain most of the chymosin. This liquid is concentrated by ultrafiltration resulting in a stock solution that is the basic material for the formulation of the final product. After the final formulation a sterile filtration step is carried out in order to remove any micro-organisms that might have been introduced during the recovery or formulation.

A comparison of the processes based on *E. coli* and *K. lactis*

The process based on *E. coli* differs from the process based on *K. lactis* in several respects. We begin the comparison by checking that you have understood the basic stages in chymosin production. Answer SAQ 5.5 before reading on.

SAQ 5.5

The diagram below represents a flow diagram of the process to produce chymosin by fermentation. Using the list given, fit in appropriate operations, inputs and outputs at the marked (*) positions to complete the diagram.

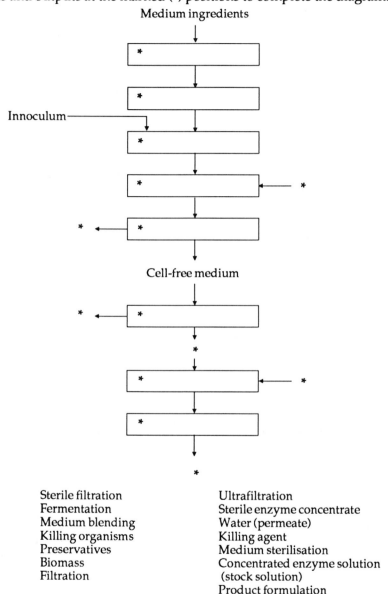

Sterile filtration
Fermentation
Medium blending
Killing organisms
Preservatives
Biomass
Filtration

Ultrafiltration
Sterile enzyme concentrate
Water (permeate)
Killing agent
Medium sterilisation
Concentrated enzyme solution
  (stock solution)
Product formulation

The fermentation medium

Obviously the medium demands will be quite different for the different organisms. Since *E. coli* is not commonly used as a production organism for food enzymes, development of a suitable industrial medium would be needed. In contrast, there is extensive experience of large scale production of *K. lactis* for food production. Because the chymosin is produced in inactive form in inclusion bodies (in *E. coli*), the effect of medium composition on the stability of the product during the fermentation seems less relevant.

The fermentation protocol

Again the lack of experience of large scale production with *E. coli* strains would call for extensive development work. Growth characteristics and regulation of the enzyme production and therefore the fermentation protocol will be quite different for the two organisms. Although no published data are available, it is very likely that the amount of enzyme per volume of fermentation liquid would be much higher with the bacterium than with the yeast. Enzyme concentration of 15% to 20% of cellular protein is common in this type of organism.

The killing procedure

Because chymosin is produced in inclusion bodies, complete autolysis would be necessary for isolation of the product from *E. coli*.

Activation of chymosin

need to re-fold product

In contrast to the process with *K. lactis*, in *E. coli* the chymosin is not directly produced in the natural conformation. Methods of completely solubilising the enzyme and refolding it in reducing conditions are technical feasible, but are too expensive for large-scale production unless the product is of very high commercial value. (It has for example been used in insulin production). It was found that a part of the product in the inclusion bodies is present in a so-called 'quasi-native' form. These molecules are in the correct folded state, but differ from the native form by the absence of cystine disulphide links. The recovery process would be designed for the maximum recovery of this 'quasi-native' product, which is then directly treated with reducing agents for conversion to native form.

Product isolation

Due to the need for complete lysis of the cells and the activation step, the liquid from which the product has to be isolated is less pure than that in the process with *K. lactis*. Therefore, the combination of filtration and concentration steps will not be sufficient to meet the desired product standards. An additional purification, such as ion-exchange chromatography, would be necessary.

5.4.4 Product formulation and specification

quality standards

The process described in the proceeding sections resulted in a liquid containing the desired chymosin. However, this was not yet a commercial product. Based on the business philosophy of the company, more detailed standards for the desired rennet product had to be defined.

market requirements

technological feasibility

Usually, the standards for the commercial product result from a combination of the marketing requirements and the technological possibilities. For the new rennet product, the existing calf rennet product was the obvious reference. Based on the market demand, the new product had to be positioned as a high quality rennet. This was in line with the expectations based on the fact that the product consists of pure chymosin. In general terms the following aspects have to be considered with regard to the quality standards.

Level of enzyme activity

The enzyme activity of existing rennets is quite different for the various suppliers and geographical areas. For the new product the activity level could easily be adapted by concentration or dilution of the stock product. In practice a product line with different activity levels is used, to meet the various market demands.

Stability

stabilisers   The activity of the product should be stable throughout the distribution period. It is a well known empirical fact that enzyme solutions are often stabilised by 'impurities' in the solution. These stabilising impurities are often minerals or biological polymers. In the extract of calf stomachs. Usually sufficient stabilising compounds are present in the extracts of calf stomachs. The fermentative production gives a purer product. Therefore special care had to be taken in the choice of the medium composition and the final product formulation to get the desired stability. On the other hand there is more freedom in the final composition, which allows for the development of extra-stable products if desired.

Microbial specification

preservation   The specification for microbial contamination of the new products had to be at least as severe as that for existing rennets. In practice, the levels of contamination (based on plate counts) for the chymosin produced by fermentation are far lower than the standards set for calf rennets, due to easier control of the production process. To maintain the low levels of contamination throughout the distribution period, the product contains a preservative such as benzoate and compounds to decrease the water activity.

Appearance

The most common rennet products are clear liquids with a brown to yellow colour. Depending on the fermentation medium and the recovery process this will be the same for the products produced by fermentation. Where additional purification, such as ion-exchange has been used, the final product may be almost colourless.

Contamination with undesired compounds

absence of DNA   A number of standards concerning heavy metals, mycotoxins, antibacterial activity and such like have been defined for all microbial food enzymes. For rennets produced with the aid of recombinant DNA technology some additional criteria had to be checked. The most important one was the absence of any genetic information that could be transferred to other organisms. This was achieved by a combination of measures. In the process based on *K. lactis*, the cell wall of the production organism was not disrupted during the killing. In combination with the design of the filtration step, this resulted in a very low DNA content in the filtrate. The conditions in subsequent stages in down-stream processing caused the breakdown of any traces of DNA that might have been present in the liquid. In particular low pH values cause depurination and hydrolysis of DNA and thus destroy the genetic information. This disruption of the DNA is even more important in the case of intracellular production of chymosin, because of the need for complete cell lysis. It was demonstrated that in the product from the process based on *E. coli*, no detectable DNA or RNA fragments larger than 132 000 kD (equivalent to 200 base pairs) were present. Other special criteria regarding contamination with undesired compounds are linked to the choice of the host organism,

toxins   for example with *E. coli* absence of toxins and β-lactamase.

∏ Why do you think the presence of β-lactamases in the product is considered undesirable? (If we tell you that penicillin is a β-lactam antibiotic, this might help you formulate an answer).

β-lactamase enzyme inactivates β-lactam antibiotics (penicillins and related antibiotics). Ingestion of such enzymes in food could reduce the effectiveness of any treatment in which such antibiotics are administered by mouth.

In order to facilitate the introduction of the product in the rather conservative dairy industry, the composition of the product was chosen to be as close as possible to that of calf rennets. This resulted in a very simple process for preparing the final formulation from the bulk product. The water activity was decreased to the desired value by addition of sodium chloride, and sodium benzoate was used as the preservative. The pH value was standardised to the optimal value for product stability and an additional sterile filtration step was performed.

5.4.5 Regulatory issues

EC directives

Before we move on to examine the features of the scale-up phases, let us examine some of the regulatory issues that are involved. It is not our intention to cover all of the many regulatory issues concerning the manufacture of food. Instead, we will focus on the more recently introduced regulations which have arisen from the application of biotechnological techniques. These regulations particularly relate to the use of genetically modified micro-organisms and their products.

We have provided two sets of information to help you understand the conditions imposed by the regulations on the use of recombinant micro-organisms. First we have provided, in Appendix 1 to this volume, a summary in the form of a commentary on two EC directives, namely:

• the contained use of genetically modified micro-organisms;

• the deliberate release of genetically modified organisms.

You may or may not have met with the contents of these directives or the regulations which arise from them in another context. It is however important that you are familiar with them.

If you can answer the following questions then you probably have sufficient knowledge of the directives and regulations, and we suggest that you read on. If you are unable to answer the questions or you are not confident of your answers, we suggest that you read Appendix 1 before reading on.

EC Member States have brought into force laws, regulations and administrative provisions necessary to comply with such EC directives. It is not our intention here to examine all of the national regulations. Instead, we have provided a summary of the regulatory position of chymosin produced from genetically engineered micro-organisms as Appendix 2. By reading this appendix you will learn much more of how the EC directives have been interpreted nationally and how the regulations have been applied to the research and marketing of recombinant chymosin. You will notice that although national regulations have much in common, there are some differences. You will notice this especially in terms of nomenclature.

We would suggest you read Appendix 2 before moving on to section 5.4.6 since this will put the development of the scale-up of the process and subsequent marketing of the product into its regulatory context.

If you have read Appendix 2 you should be able to answer the following questions.

SAQ 5.7

1) What are the three main regulatory regimes that the production of chymosin are subject to?

2) What is the difference in the category nomenclature given to *K. lactis* in The Netherlands and France? What would its categorization have been in the UK?

3) What are the three basic principles which should be met before a food is placed on the market?

4) To gain authorisation to market a product, it has to be shown to be safe. This is done by producing a safety dossier. Give a list of contents of what such a dossier should contain?

SAQ 5.6a

1) What is meant by a GMMO or a GMO?

2) Which of the following techniques for genetically modifying a micro-organism are exempt from the EC directive on the contained use of genetically modified micro-organisms: cell fusion by artificial means, electroporation, mutagenesis, self-cloning in some non-pathogenic organisms, somatic animal cell hybridisation?

Now let us return to the development of the recombinant chymosin production process. We now focus on the scale-up of the process.

5.4.6 Scale-up

At the end of the research phase of the project, a strain capable of producing chymosin was available and the biochemical identity of the enzyme had been established. Furthermore, a process for the production of the final product according to the defined criteria had been designed on the basis of laboratory-scale experiments. The next step was to check whether the process could be carried out on a larger scale and whether this product met the criteria derived from the demands of legislative procedures and application in cheese manufacturing.

Safety assessment

Research with recombinant DNA micro-organisms is restricted by a number of regulations. These rules allow for fermentation experiments up to a volume of ten litres (ie a Type A process - see Appendix 1). In order to get an approval for production of larger volumes, (ie a Type B process -see Appendix 1) each process has to be agreed with the responsible authorities.

SAQ 5.6b

3) To which group should a pathogenic strain of *Salmonella typhi* be assigned? (Group I or Group II)?

4) What is a Type B operation?

5) The EC directive on the deliberate release of genetically modified organisms demands that a notification must be submitted. To whom must it be submitted and what should it contain?

6) When placing a product containing genetically modified micro-organisms on the market, who should be informed?

We have learnt in the previous section that the process with a chymosin-producing *K. lactis* strain, the least stringent level of containment (GILSP, Good Industrial Large Scale Production) has been accepted.

checking of process

survival of released organisms

Before scale-up trials were carried out, the characteristics of the strain and the enzyme produced had to be studied at the laboratory scale. In particular the stability of the recombinant DNA expression cassette had to be proven. Also the protocol for killing the production organism in the fermenter had to be checked for each strain that was to be tried on a larger scale. Furthermore, the large-scale protocol and equipment have to be suitable for use with recombinant strains. As part of the evaluation of potential environmental hazards, the survival of the *K. lactis* production strain in a number of soil and water samples was compared to that of the host strain. Depending on the conditions, the number of living cells was reduced to zero in 24-36 hours. This rapid elimination of the micro-organism in the environment is an important form of containment.

A pathogenicity study was performed on the chymosin-producing *K. lactis* strain. The results of this study showed that the recombinant strain was non-pathogenic.

∏ Make a list of tests you would perform to see if the organism was pathogenic.

You obviously cannot start off by injecting the organism into humans. Your list of tests should have included injecting laboratory animals such as mice by various routes with various doses of the organism, and observed to see if any infection results.

Scale-up of the production process

When the conditions for large-scale culture had been met, the process that had been developed on a laboratory scale was transferred to the pilot plant. In the scale-up of the process based on *K. lactis* no special problems were experienced. In view of the fact that the process was based on the experience with this organism in large-scale production of lactase, this was as expected.

The product from the pilot plant was used to evaluate its compliance to the standards defined by registration and marketing strategies (we have dealt with this in Appendix 2).

Toxicological testing

Apart from the evidence for non-pathogenicity of the production organism, the non-toxicity of the final product also had to be guaranteed. For the different aspects of toxicity there were numerous test methods available. As a result of dialogue with the authorities in various countries a set of tests was selected, which together supplied the desired evidence.

∏   Make a list of the tests you think should be carried out on chymosin. When you have done this compare it with the list we have provided in Appendix 2 and with the discussion below.

mutagenicity

AMES test

acute oral toxicity

The absence of mutagenicity was checked using the AMES test. This test basically looks at mutation rates in certain bacteria. Acute oral toxicity was examined in a limit test, a method advocated for determining the acute toxicity of compounds with expected low toxicity. A single dose was administered orally to rats. During the 14 days observation period, no abnormalities were detected. It was concluded that the product exhibits a low order of toxicity.

feeding trials

To check whether repeated ingestion of cheese prepared with chymosin produced by *K. lactis* would produce hazard to health, three groups of rats were observed for 91 days. The test group received cheese produced with the biotechnologically produced chymosin and the control group the same amount of cheese produced with traditional rennet. The third group received only the standard diet. With the exception of a growth increase in both groups receiving cheese, no treatment-related effects occurred. An even more stringent test was the oral administration of the chymosin product directly to the rats. Again the results showed that the biotechnologically produced chymosin did not cause toxicity.

allergenicity

The allergenicity of the product was tested on guinea pigs by application to the skin. Results were compared with those obtained by testing a commercial calf rennet product. It was concluded that the recombinant chymosin product was not a strong skin-sensitising agent.

Functional properties of the biotechnologically produced chymosin

The enzymatic activity of the chymosin from *K. lactis* was compared with a standard animal rennet (INRA-Poligny/France) in several respects as follows.

- Effect of enzyme concentration

comparable kinetic behaviour

The effect of the enzyme concentration on the coagulation time of raw milk was determined for both rennets. The same reciprocal linear relation was found for the two preparations, which indicates that both products show the same kinetic behaviour.

- Effect of temperature and pH

For both parameters, the same relation was found for the two enzyme products. The temperature optimum was reached at 45°C. Above 55°C and below 15°C the enzymes were inactive.

In the range between pH 6.0 and pH 6.8, both rennets possessed the same sensitivity to pH value. For the enzyme produced by *K. lactis*, the pH optimum was determined to be 5.8.

- Effect of calcium concentration

The stimulating effect of calcium on the coagulation time was tested for the two rennets. It was found that at pH 6.5 the animal rennet was stimulated slightly more than the chymosin from *K. lactis* in particular in the case of raw milk. At pH 7.0 this difference did not occur. This phenomenon is explained by the greater stimulation of the coagulation catalysed by pepsin as compared to chymosin. In the determination of the activity of commercial products, 10 mM calcium is added, which corresponds to the maximum value used in cheese making.

- Inactivation in whey

The inactivation of both rennets during pasteurisation of whey (72°C, 15 sec) at different pH values was identical.

- Gelation of the curd

The kinetics of the gel-forming process were followed for both enzyme preparations. When coagulation times were adjusted, no differences in gel formation were found.

Cheese manufacturing trials.

The main criteria determining the quality of a rennet are those concerning its behaviour in the cheese making process. Cheese is the collective name for a wide variety of products. The manufacturing processes employed for the different types of cheese each have their specific characteristics and the criteria for the assessment of quality of cheese

local preferences

may differ by type. These aspects were examined in Chapter 4. Furthermore, local preferences will influence the judgement of cheese quality.

local comparison

In the evaluation of the chymosin produced from *K.lactis*, the experimental strategy was to perform trials in different countries on the most typical types of cheese. It was assumed that the judgement on the quality of cheese would be most critical in the region of origin. As a reference, a locally available good-quality calf rennet product was used. Taken together these trials gave a good overall image of the quality of the new type of rennet.

The rates of coagulation of the milk, the strength of gels formed and syneresis with the chymosin derived from *K. lactis* were equal to the values obtained with the reference rennets. Some non-significant deviations were seen, which could easily be compensated by small adaptations in the process conditions. Panels of volunteers were used to test

organoleptic

the organoleptic properties of the cheeses - that is the taste, smell and texture (or what is referred to as 'mouth feel'). It was concluded that no overall differences were apparent between cheese made with the different rennets. The conclusion drawn from the various comparative cheese manufacturing trials was that with the use of chymosin produced with recombinant *K. lactis* results were equivalent to those obtained with traditional rennet.

5.4.7 Commercial operation

At the end of the scale-up phase it had been established that the desired product could be produced on a commercial scale. In other words, the product was ready for introduction in the market. For the major part, the activities in the commercial phase of the project concerning chymosin production by a recombinant DNA micro-organism, were not essentially different from those for any other food enzyme. These production and marketing activities will not be discussed in this case study.

influence on
legislation

However, the pioneering aspects of chymosin as a food enzyme produced via a genetically engineered micro-organism, did have an influence on the legislation procedures. The standards for the evaluation of the safety of the product differed from country to country. To overcome this, the guidelines of the AMFEP (Association of Microbial Food Enzyme Producers) were used as a basis. The dossier for regulatory review was then adapted to the specific requirements of the various countries. As discussed earlier, special attention was paid to minimise the potential risk of DNA fragments in the final product. This was done by maintaining the concentration at a low level and by creating conditions that cause the breakdown of any traces of DNA.

additional
standards

Once the absence of any hazard to consumers' health had been proven to the satisfaction of the responsible authorities, the new product could be accepted for application in food. Although this was a pre-requisite for use in cheese production, in most countries additional standards had to be met. For certain food products special standards exist, which specify in detail the ingredients that are allowed for these products. Depending upon the exact description of the coagulation agents, the new product may or may not fit in with such specifications.

acceptance by
industry

In addition to these legal aspects, the acceptance of the new product by the dairy industry had to be taken into account. As the attitude of the dairy industry is traditional, this was usually the most difficult part. One obvious way to facilitate the acceptance was to make as much information as possible available to the industry. Furthermore, the benefits (on cheese quality and/or production costs) have to be significant. Once these benefits are recognised, it will only be a matter of time before this example of advanced biotechnological development will represent an attractive commercial product for the dairy industry.

SAQ 5.8

Choose the correct completion to the statement.

Chymosin production using genetically engineered microorganisms was developed because:

1) this produces a superior product compared to calf rennet;

2) not enough rennet to supply the needs of the cheese industry could be made from calves' stomachs;

3) fermentation processes can produce enzymes on a large scale;

4) the technology to produce such enzymes by fermentation already existed.

SAQ 5.9	Listed below are some of the steps in the development of the procedure to produce chymosin by a fermentation process. List these steps in the order that you think they would be performed, starting with the first and ending with the last. If you think any steps would be performed concurrently (at the same time) list them side by side.

Scale up the fermentation process.

Introduce heterologous DNA into host organism.

Identify product of expressed heterologous DNA in host organism.

Conduct manufacturing trials.

Develop a method of killing the organism in the fermentation medium.

Develop a fermentation medium.

Commercial operation.

Select host organism.

Summary and objectives

Chymosin is not a clear cut example of an additive or food ingredient. This means that the regulatory position is not quite clear and in many cases more than one set of regulations is applicable. It has turned out that a reasonable approach to the safety questions guided by common sense rather than regulatory requirements is quite acceptable to authorities. Openness and willingness to discuss relevant aspects has been very much appreciated.

The influence of the recombinant DNA aspect in the regulatory process was noted but certainly not regarded as a negative issue. From the experience obtained so far we may conclude that existing regulations can encompass these products.

On the other hand it has also become clear that existing regulations in this area within the EC are widely different. A common approach in the Member States will certainly improve opportunities for industry to introduce innovative products.

Now that you have completed this chapter you should be able to:

* describe the role of rennet in cheese manufacture;

* relate the mode of action of rennet to the coagulation process;

* describe the sources and activities of rennet and rennet substitutes;

* describe the comparative advantages and disadvantages of rennet and rennet substitutes;

* explain the rationale for producing chymosin by genetic engineering;

* describe how a production process for recombinant chymosin was developed from inception to production;

* draw and interpret a flow diagram showing the production of chymosin from a recombinant micro-organism by a fermentation process.

* outline the legislative and cultural constraints placed upon the introduction of rennet substitutes.

High fructose corn syrups

High fructose corn syrups

6.1 Introduction

Those of you with a 'sweet tooth' will appreciate the liking many humans have for sweet tasting foods. Sugars are common constituents of many foods: sucrose, fructose, glucose and many others in fruits, plants and vegetables; lactose in milk. The sugars tend to impart a pleasant sweetness upon foods which contain them. This has led humans to produce sweeteners, which can then be used to impart the same pleasant sweetness upon foods which are not sweet by nature. The traditional sweetener in most countries is sucrose. Sucrose is abundant in certain crops such as sugar cane and sugar beet, and for well over two centuries the trade in sugar has been international. However, developments have taken place to produce alternative and/or superior sweeteners to sucrose. We are going to examine these developments in this chapter, with emphasis on the production of high fructose corn syrups. In this chapter corn refers to what Americans call corn (ie maize) rather than another type of grain.

6.2 Important terms used in the glucose/fructose industry

The glucose and fructose industry has its own nomenclature. Before you read the chapter, familiarise yourself with the following items.

Amylopectin - A complex branched molecule consisting of amylose chains linked at branch points by α-1,6 links. Corn starch contains about 74% w/w of amylopectin. The average degree of polymerisation of amylopectin molecules is about 2 million.

Amylose - A glucose polymer consisting of linear chains of glucose, linked by α-1,4 linkages. Corn starch contains about 26% w/w of amylose. Corn starch amylose has an average degree of polymerisation of approx 800 (ie it contains about 800 glucose molecules), while the degree of polymerisation of potato starch amylose reaches approx 4900.

Caramelisation - The oxidation of sugars upon heating. When heated, a sugar solution turns brown and acquires a characteristic flavour. Caramel is a strongly heated sugar solution, and is widely used as a colouring and flavouring agent (for examples in colas).

DE - (Dextrose equivalent) refers to the measurement of the total reducing sugars in a syrup, calculated as dextrose (D-glucose) and expressed as a percentage by weight of the total dry substance. Pure dextrose has a DE of 100, starch has a DE of 0. The greater the DE value (of a starch solution), the greater the degree of hydrolysis of the starch.

Degree of Polymerisation (DP) - The number of molecules in a polymer or oligomer.

Dextrinisation - The breakdown of starch into short oligomers of glucose.

Dextrose - Dextrose is crystallised D-glucose. Dextrose can be obtained in either anhydrous or monohydrate form.

Dry solids - The carbohydrate material which remains when the water is evaporated from a starch or sugar solution.

Gelatinisation temperature - The temperature at which the starch granules swell, the starch polymers in the granule are hydrated and the crystalline structure of the granule is lost.

HFCS - High fructose corn syrup. This describes the standard 42% (as a % of total sugars) fructose syrup made from corn. Syrups containing 55% of fructose are given many names like ultra high fructose glucose syrup (UHFGS) or second generation HFCS.

Isomerisation - The enzymatic conversion of glucose into fructose.

Laevulose - An alternative name for fructose.

Liquefaction - The first stage of starch cooking in which the starch grain is disrupted and the starch rendered soluble. This process can be done either with acids or with enzymes.

Retrogradation - The process in which amylose molecules spontaneously become oriented into semi-crystalline arrays which are disrupted only with high mechanical forces with great difficulties. It occurs when very low DE material is allowed to cool.

Saccharification- The process in which liquefied starch (DE 10-20) is hydrolysed by acid or an enzyme such as glucoamylase to produce glucose.

Starch - A mixture of two glucose polymers - amylose and amylopectin.

Steeping - Soaking of the corn in warm water to soften the kernel and to extract the majority of the carbohydrates, minerals and protein.

SAQ 6.1

As a check of whether or not you have understood these terms, answer the following:

1) Name a polymer of glucose which contains some α-1,6 links.

2) If starch is partially hydrolyzed, will its DE value rise or fall?

3) Does the DP value of starch rise or fall during saccharification?

4) Is the process by which starch granules take up water to form a gel rather than a crystalline structure called retrogradation, isomerisation, caramelisation or some other term?

5) Is it true that the complete hydrolysis of a polymer of fructose would produce laevulose?

6.3 Sugar and the sugar industry

We have already seen in Chapter 2 how consumption of sugar (sucrose) has declined over the past two decades in favour of other sweeteners. The search for alternative sweeteners goes back further than this. The Napoleonic wars in Europe in the early 1800s isolated much of Europe from traditional sources of sugar in the West Indies and elsewhere and motivated a search for alternatives.

saccharification

In 1811 a German chemist, Kirchhoff, applied acid to the saccharification of starch, and by 1850 some large-scale acid saccharification processes were in operation. However the introduction of sugar beet into European farming boosted sugar production and it was not until the Second World War that saccharification again became important, and enzymatic saccharification processes were developed.

fluctuation of sugar prices

In the 1960s, the sugar price on world markets changed unpredictably, resulting from factors such as the revolution in Cuba (a major sugar exporter). Prices rose over the late 1960s, until in 1973 sugar was priced at US $0.24 kg^{-1}. In 1974 the world market again became unstable and prices soared to an all-time high of US $1.30 kg^{-1}. The 1990 sugar price stands at about US $0.30 kg^{-1}.

search for alternative sweeteners

High sugar prices caused the food industry to look carefully at alternative sweeteners. Sweeteners are used widely in foods. Apart from the cost considerations, the sugar industry was interested in alternative carbohydrate sweeteners on other accounts. Although sucrose is relatively sweet compared to other common sugars (except fructose - see Table 6.1) it does have lower solubility than, for example, glucose and fructose. This can lead to crystallisation of sucrose upon cooling/chilling of foods.

Substance (crystalline form)	Relative sweetness
sucrose/saccharose	1.0
lactose	0.4
glucose/dextrose	0.7
fructose	1.3
mannitol	0.7
sorbitol	0.5
xylitol	1.0
glucose syrup 42 DE	0.4
glucose syrup 64 DE	0.6
42% HFCS	1.0
55% HFCS	1.1

Table 6.1 Sweetness of various carbohydrate sweeteners.

∏   What do you think the problems of crystallisation of sucrose are in foods?

You probably like ice cream and you probably like soft drinks. Would you like ice cream with hard crystals in it, or crystals in your favourite cola? Many prepared sweet foods are preserved by deep freezing and refrigeration. Crystallisation can cause unpleasant 'mouthfeel' in such foods.

For these reasons, in the early 1970s, there was much interest in producing alternative carbohydrate sweeteners from starch. The technology to do this using microbial enzymes was well developed and large and stable supplies of starch were available in both Europe and North America.

∏ Make a list of the major sources of starch and compare it with our list provided below.

Grains are rich in starch. In North America, corn (maize) is produced in large quantities at low cost. In Europe wheat and other grains are available. At present, about 20×10^6 tons of starch are produced annually world wide, and about 75% of this is derived from corn. The advantage of grains, compared to other starchy crops such as potatoes, is that they are easy to store. The 1990 price of maize was about $0.08 kg^{-1}$.

acid hydrolysis and bitter side products

Although starch can be hydrolysed by acid to produce syrups with high DE, this process is accompanied by the production of bitter by-products. Enzymes are specific in their activity, operative at low temperatures and have been applied to starch hydrolysis without formation of bitter by-products. The ideal product from starch would be glucose (ie complete hydrolysis).

advantages of using enzymic hydrolysis

We will now examine how enzymes are applied to the production of sweeteners from starch. To do this we will first need to learn something of the structure of starch. Then we will examine the stages in its conversion to sweeteners (starch liquefaction; dextrin formation; high glucose syrup production; HFCS production).

6.4 The structure of starch

Starch is a major component of many grains, and other crops (see Table 6.2). There are regional variations in the sources of starch available, for instance in the USA corn is used, in Europe wheat or potatoes, in Asia, rice.

∏ Which of the starch sources listed in Table 6.2 has a) a distinctive temperature of gelatinisation and b) a particularly high proportion of amylopectin?

The gelatinisation temperature of barley starch is lower than the other starches listed. Potato starch has the highest proportion of amylopectin.

Starch source	Content (%dry basis)		Compostion of starch (%)		Gelatinisation temperature
	starch	protein	amylose	amylopectin	°C
corn (maize)	72	10	27	73	62-71
sorghum	74	12	28	72	67-75
wheat	76	12	26	74	68-79
barley	64	13	22	78	54-63
potato	75	8	20	80	70

Table 6.2 Composition of starch sources.

amylose

amylopectin

Starch is composed of two types of polysaccharide chains of D-glucose called amylose and amylopectin. The ratio of amylose : amylopectin varies from species to species (see Table 6.2). The structures of these two polymers are depicted in Figure 6.1.

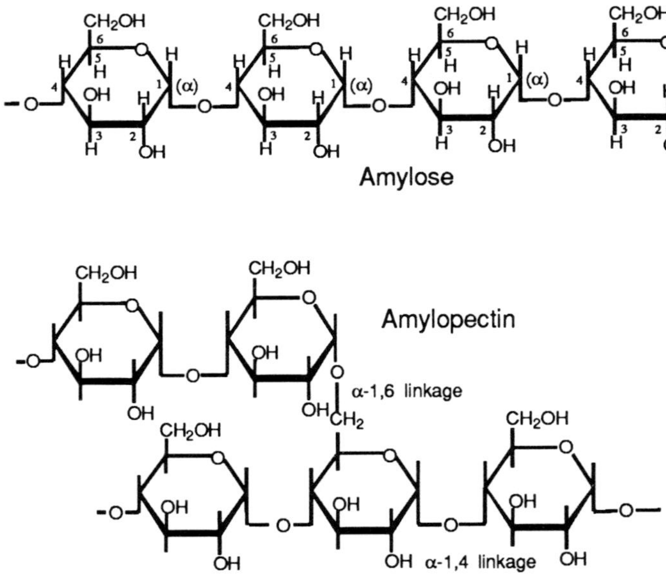

Figure 6.1 Starch structure: amylose and amylopectin. Amylose is a linear polymer of α-1,4 linked D-glucose with DP around 6000. Amylopectin is a highly branched polymer of short amylose chains connected by α-1,6 linkages and with DP around 106.

In plant material, starch is present in dehydrated form and as relatively large granules.

∏ Are starch granules from plant materials soluble in water? (Think about what happens if you put peeled potatoes in a bowl of water).

The answer is no. If you put wheat, potato or rice flour in water it does not dissolve; the granules form a suspension or settle out. Starch is stored in this form in cells so that it is relatively inert, and so does not affect cellular function.

∏ How can you dissolve starch granules, and what happens to the solution when starch dissolves?

gel formation If you are a cook, you will realise that starch can be dissolved by heating. Once dissolved, the starch forms a viscous solution (gel). The viscosity of starch solutions increases as the concentration increases and decreases as the temperature increases.

6.5 Starch production: The wet-milling process

wet-milling

bran (hull)

germ

gluten

starch

The process to recover starch from corn, known as wet-milling, is essentially a method of disrupting the corn kernel in such a way that the component parts can be separated in an aqueous medium into relative pure fractions. Corn kernels (seeds) mature in about 60 days following pollination and are harvested in late summer or early fall (autumn) when kernel moisture content has dropped to below 30% (wet weight basis). In addition to water, corn consists of four major components that must be separated in the wet-milling process. These are: the outer skin, called bran or hull; the germ, which contains most of the oil; the gluten, which contains most of the protein; and the starch (see Table 6.3 for a fuller analysis).

Component	Content %
moisture	17.0
starch	70.0
protein	10.0
fat	4.5
ash	1.4
pentoglycans	6.2
crude fibre	2.7
cellulose-lignin	3.3
sugars	2.0

Table 6.3 Analysis of corn grain (% dry basis - except moisture which is % wet basis).

waxy corn hybrids Plant breeding has resulted in the development of hybrids with higher contents of oil and/or protein. The most important genetic change in corn for starch production, however, has been the development of waxy corn hybrids. The starch from waxy corn, which contains 100% of the branched starch fraction, amylopectin, has become a major food starch, mainly as a thickener in various food products.

Modern wet-milling of corn is carried out in a continuous process. The process is described in Figure 6.2.

Corn first passes through a screening (sieving) step to remove foreign material. Prior to wet-milling the corn must be softened. For this it is normally steeped for 30-40 hours at a temperature of 48-52°C. The majority of the soluble carbohydrates, minerals and proteins are liberated. Water, which is pumped through the steeping tanks continuously, is finally removed and evaporated. The resulting concentrate, which is

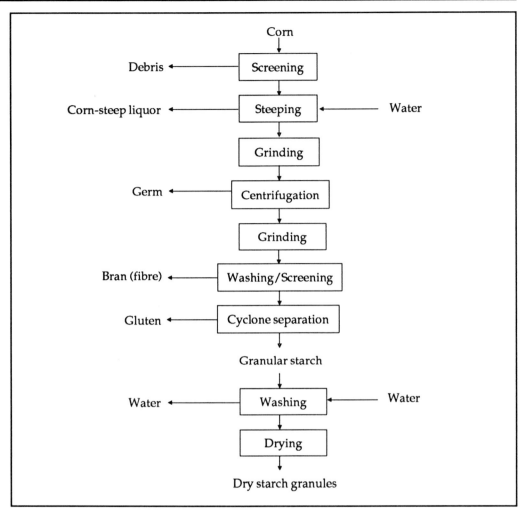

Figure 6.2 The wet-milling process.

known as corn-steep liquor, is sold as a medium ingredient for fermentation or as a feed additive.

∏ Make a list of the activities you think lactic acid bacteria might have during the steeping process. (In making your list, think back to what you learnt about lactic acid bacteria in cheese manufacture).

Lactic acid bacteria ferment some of the sugars released from the maize to lactic acid. This, as you must know if you have read the previous chapters, lowers the pH and prevents microbial growth and spoilage. They also release proteases which degrade proteins. This increases the amino acid content of corn steep liquor and prevents foam formation during its processing.

The softened corn is coarsely ground in a mill (grinder). The bulk of the germ is freed and removed by centrifugation. It is then processed for direct sale or for oil recovery.

The remaining mixture of starch, gluten, fibre and bran is then finely ground and countercurrently washed through a series of screens to separate the bran and fibre from the starch and gluten. High speed cyclone separators (Figure 6.3) are used to separate the relative heavier starch from the lighter gluten. The separated gluten is then processed to produce a dry gluten feed product.

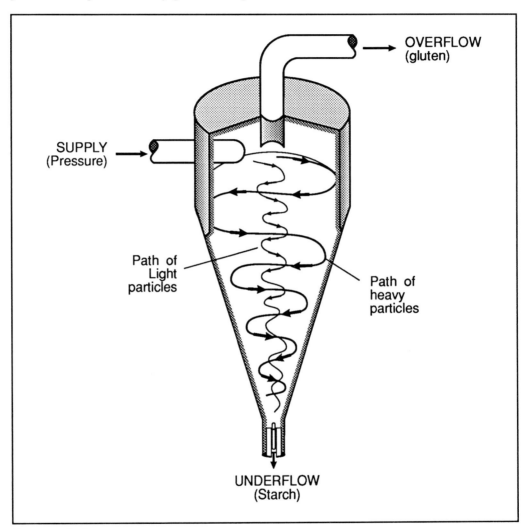

Figure 6.3 Cyclone separater.

The starch stream from the centrifugal separation then proceeds to the final step in the wet-milling process, multistage starch washing, which is designed to remove any gluten and solubles remaining in the starch stream. After proper washing, the granular starch typically contains less than 0.3% total protein and less than 0.01% soluble protein. The starch granules are then dried, to allow storage.

6.6 From starch to dextrins

6.6.1 Objectives

What would be the ideal achievement for the enzymatic conversion of starch to dextrose? The goal would be to reach total conversion to dextrose or, using the terminology of the field, DE 100, and to do this in a cost effective way. In the following section we will examine approaches to achieving these objectives.

6.6.2 The ideal system for starch digestion

Let us consider what characteristics the ideal system for the enzymatic degradation of starch to glucose would have. For ease of operation and effectiveness the system should have the following characteristics:

- the enzyme(s) should be active on native starch (granular corn starch as it is produced by wet-milling);

- degradation should take place in one step (and so involve only one enzyme);

- degradation should take place at about pH 5 (the pH of the granules) and at low temperatures;

- degradation should be complete (and lead to a material with DE value of 100);

- the enzyme should be active when immobilised, so that continuous systems can be operated;

- the enzyme should be easy and cheap to produce.

∏ Examine each of these objectives in turn and see if you can see any obvious difficulties in achieving any of them in practice. (Then read about the difficulties we can envisage).

liquefaction

We have seen already that starch granules are relatively large masses of insoluble dehydrated starch material. You would not expect enzymes to be highly active on these, even if the granules were suspended in water. Such granules form a slurry when mixed in water, which would make operation of continuous systems difficult (the granules would clog the support matrix for the immobilised enzyme). In addition, a single enzyme is not available which is capable of hydrolysing granular corn starch to glucose (a process called saccharification). Instead, granular starch must first be converted to soluble form (a process called liquefaction). Let us now examine these processes in detail.

6.7 Starch liquefaction

jet cooked

gelatinisation
dextrins

dextrinisation

Starch is added to water to produce a slurry containing 30-40% (w/v) dry solids. The pH is raised to 6.0-6.5, α-amylase and $CaCl_2$ (enzyme stabiliser) are added and the mixture is heated (jet cooked) at 105-110°C for 3-7 minutes. Sometimes, jet cooking at 140°C, in the absence of enzymes, is applied as a pre-treatment step. During this time the starch granules absorb water, swell and finally disintegrate. This is called gelatinisation. Once the granules have gelatinised, the temperature is lowered to 95-97°C, more α-amylase may be added, and the mixture is held for 60-90 minutes. During this time, the gelatinised starch is degraded to short glucose chains (degree of polymerisation 2-7) called dextrins. Because of this, the process is called dextrinisation. The product of dextrinisation is a viscous dextrin syrup (DE 10-12), in which the insoluble starch has been converted to soluble dextrins. These dextrins are easy to convert to glucose by further enzymic hydrolysis.

SAQ 6.2

The following flow diagram represents the liquefaction process for corn starch. Fit the inputs and operations listed below into the flow diagram to complete it.

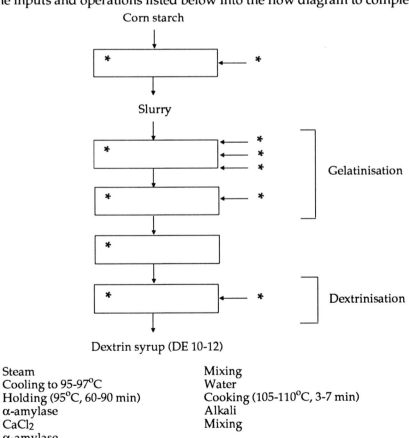

Steam
Cooling to 95-97°C
Holding (95°C, 60-90 min)
α-amylase
$CaCl_2$
α-amylase

Mixing
Water
Cooking (105-110°C, 3-7 min)
Alkali
Mixing

6.7.1 Characteristics of α-amylases used in starch liquefaction

thermostable

The α-amylase first used in starch liquefaction was a thermostable enzyme from the bacterium *Bacillus amyloliquefaciens*. The cooking temperatures of 140°C caused some inactivation of the enzyme and so an additional dose of enzyme was needed for dextrinisation.

∏ A thermostable enzyme is required because gelatinisation and dextrinisation are carried out at high temperatures. Use your knowledge of the properties of starch and starch-derived materials to explain why these processes cannot be carried out at lower temperatures.

retrogradation

Both processes result in a viscous liquid being formed, which only runs freely at relatively high temperatures. In addition, after gelatinisation, some amylose is formed. If the solution is allowed to cool, the amylose crystallises in a process called retrogradation. These crystals are extremely difficult to redissolve.

mode of action
of α-amylase

The bacterial α-amylase (EC 3.2.1.1) is an endohydrolase (acting within the molecule), which acts randomly on α-1,4 linkages in amylose and amylopectin (Figure 6.4). α-1,6 linkages and α-1,4 linkages adjacent to them are not hydrolysed. This produces a mixture of linear and branched dextrins with DP 2-7. The enzyme has lower affinity for dextrins with low DP, compared to longer oligosaccharides, and thus these are hydrolysed relatively slowly (Figure 6.5).

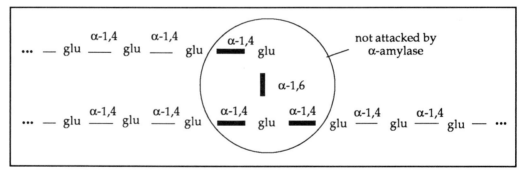

Figure 6.4 The action of α-amylase on starch. Bacterial α-amylase does not act on α-1,6 links or α-1,4 links adjacent to them (encircled). It may attack any of the linkages outside of the circle.

6.7.2 Improved enzymes

∏ Make a list of the improvement in α-amylase performance which would be useful in the liquefaction process.

The major improvement would be an enzyme which was more thermostable, and so able to survive cooking and remain operative for dextrinisation (thus overcoming the need for the second dosing). If at the same time a greater degree of dextrinisation was produced this would also be an advantage. High affinity for dextrins with DP 2-7 could also be an advantage.

For a more-thermostable enzyme, other mesophilic and also thermophilic strains of *Bacillus* and *Clostridium* were screened. Several such α-amylases were discovered, and two, from *B. licheniformis* and *B. stearothermophilus*, have been developed for industrial-scale operation.

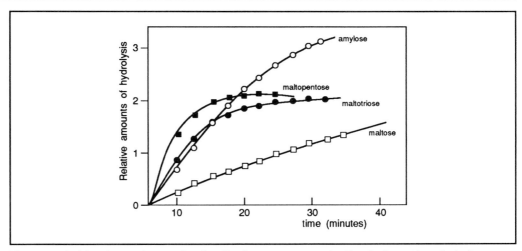

Figure 6.5 Activity of α-amylase on dextrins. For these assays standard incubation regimes including substrate concentrations were used.

6.7.3 α-Amylase from *B. licheniformis*

α-Amylase from *B. licheniformis* produces DP 5, 3, 2 and 1 products (in descending order of concentration). It does not produce significant quantities of DP 4, 6 or 7 products. This enzyme, like that from *B. amyloliquefaciens*, preferentially cleaves intermediate-length oligosaccharides close to their reducing end. The enzyme requires Ca^{2+} addition for stability and is active and stable at temperatures of 105-110°C. Consequently, gelatinisation and dextrinisation can be carried out using a single dose of enzyme. For liquefaction using enzyme from *B. amyloliquefaciens*, the enzyme requirement is 1.0-2.0 kg liquid enzyme concentrate per ton dry starch substance. For the enzyme from *B. licheniformis* the equivalent requirement is 0.5-0.7 kg.

increased stability

You may well be wondering why a mesophilic organism such as *B. licheniformis* should produce an enzyme which is stable at high temperatures (far beyond temperatures at which the organism can grow or even survive). One suggestion is that, as the organism also produces proteases, it has developed a very compact amylase molecule which is relatively insensitive to degradation by proteases. The compact configuration, coupled with stabilisation by the starch substrate, seems to confer thermostability upon the amylase enzyme.

In practice a one-step liquefaction process is not always followed. After gelatinisation the solution is sometimes heated to above 120°C to make filtration easier later on in the process. This reduces enzyme activity and so a second dose is applied for dextrinisation.

6.7.4 α-Amylase from *B. stearothermophilus*

comparison of α-amylase from B. stearo- thermophilus and other sources

α-Amylase from *B. stearothermophilus* produces DP2-DP5 products in roughly equal proportions. It is operative at lower pH than the previously mentioned α-amylases. This has several advantages for processing. It avoids the need for alkali addition (and the subsequent ion-exchange process to remove the salts which this process forms); it reduces the formation of unwanted by-products such as maltulose (α-D-glucopyranosyl-D-fructose) which is resistant to further enzymatic degradation and thus reduces the DE that can be achieved.

∏ What is the next logical step after the production of dextrins in the process of producing a sweetener from starch?

Having produced a solution of linear and branched dextrins, the next step is to hydrolyse these to glucose. Thus the syrup with DE 10-12 is converted to a syrup with DE value of approximately 100 (glucose syrup). Let us examine how enzymes are applied to this.

6.8 Saccharification of dextrin syrups

saccharification

Dextrin syrups can be further hydrolysed to constituent glucose by enzyme action. Such hydrolysis is called saccharification. It has been recognised since the 1950s that fungi such as *Aspergillus spp.* and *Rhizopus spp.* produce enzymes capable of such activity.

exohydrolase

The enzyme amyloglucosidase (or glucoamylase) EC 3.2.1.3 is an exohydrolase which acts on gelatinised starch or dextrins by removing single glucose molecules from the non-reducing ends in a stepwise fashion. The enzyme acts primarily on α-1,4 glucosidic bonds but it will also slowly hydrolyse α-1,6 bonds.

6.8.1 Characteristics of fungal amyloglucosidase

synergism between α-amylase and amylo-glucosidase

The kinetics of dextrin hydrolysis by amyloglucosidase is complicated by the fact that the enzyme is acting on oligosaccharides of different chain lengths, and the chain length has an effect upon the rate of hydrolysis. It is at a maximum with chain lengths of 4-5 (ie substrates maltatetraose and maltapentaose). Thus, α-amylase and amyloglucosidase operate synergistically (the combined effect is greater than the sum of their individual effects): α-amylase can produce oligosaccharides of chain length 4-5, upon which amyloglucosidase acts most rapidly.

relative thermo-instability of amylo-glucosidase

Amyloglucosidase is not as thermostable as α-amylase. At room temperatures the enzyme activity is stable over a wide pH range (3-9), but at 55°C the enzyme is active over a narrower range. Amyloglucosidase is a glycoprotein, containing 5-30% carbohydrate. This partly accounts for its decreased thermostability compared to α-amylase. The enzyme is not stabilised by Ca^{2+} ions.

The influence of pH on enzyme activity and stability are shown in Figures 6.6 and 6.7 respectively. The influence of temperature on enzyme activity and stability are shown in Figures 6.8 and 6.9.

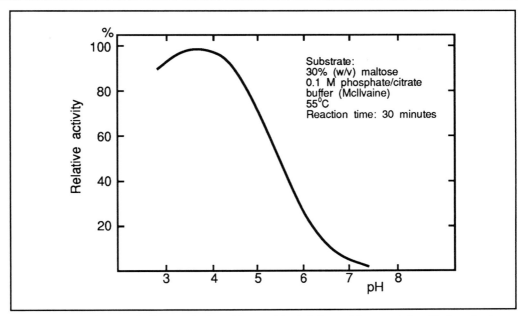

Figure 6.6 The influence of pH on amyloglucosidase activity (relative activity reported as % of maximal activity).

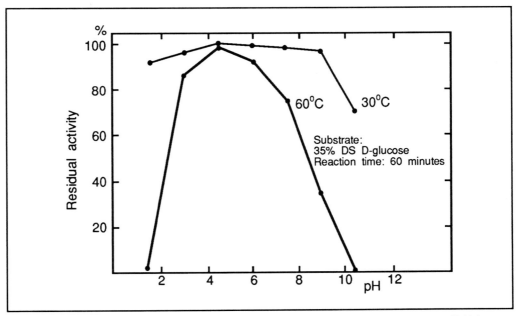

Figure 6.7 The influence of pH on amyloglucosidase stability (residual activity reported as % of initial activity).

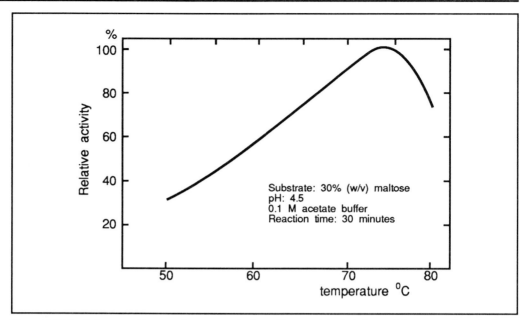

Figure 6.8 The influence of temperature on amyloglucosidase activity (relative activity reported as % of maximum activity).

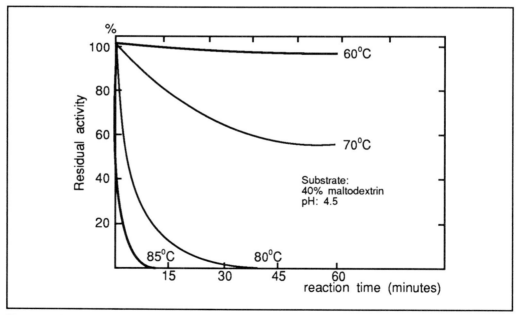

Figure 6.9 The influence of temperature on amyloglucosidase stability (residual activity reported as % of mitial activity).

∏ What do you think would be the optimum working conditions for amyloglucosidase, with respect to pH and temperature.

The optimum working conditions would be pH 4.0-4.5 (Figures 6.6 and 6.7). Although the optimum activity of the enzyme occurs at about 75°C (Figure 6.8), at this temperature the enzyme is very unstable (Figure 6.9). In practice operating temperatures of around 60°C are used.

The amyloglucosidase preparations from *Aspergillus niger* have been shown to contain at least three amyloglucosidase isoenzymes and also an α-amylase (called acid α-amylase). It is important the preparation is free of transglucosidase, as this enzyme causes formation of isomaltose (a dimer of α-1,6 linked glucose), which is only partly broken down to glucose by amyloglucosidase. This lowers the yield of glucose.

transglucosidase

The hydrolysis of dextrins by the three different isoenzymes of amyloglucosidase, under industrial working conditions (33% dry solids, pH 4.2, 60°C) is shown in Figure 6.10.

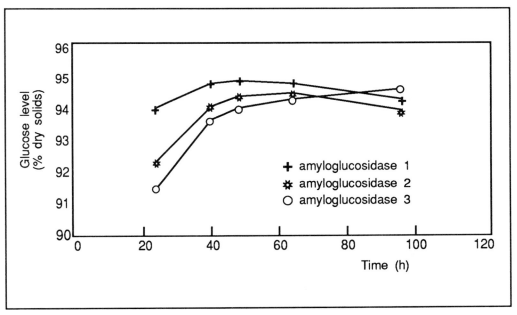

Figure 6.10 Saccharification by a number of amyloglucosidase isoenzymes.

∏ Use the data in Figure 6.10 to answer the following questions. 1) When is the rate of glucose production most rapid? 2) When is maximum amount of glucose formed? 3) What is the greatest % of glucose produced?

Our responses are:

1) The rate of glucose formation is most rapid in initial stages of hydrolysis. The rate decreases due to the lower activity of amyloglucosidase on α-1,6 linkages.

2) Glucose production reaches maximum levels at about 45-50 hours. After this period it can decline due to isomaltose formation.

3) The highest % of glucose formed is 95. This level could be increased if the system was operated with lower concentrations of dry solids (<35%). We will not discuss

the reasons for this here; it has to do with thermodynamics. However, lowering the substrate concentration leads to increased processing costs (the volume of material would be increased).

∏ Can you suggest what type of enzyme could be added to amyloglucosidase to improve its (amyloglucosidase) performance?

We have already seen that amyloglucosidase acts by splitting off glucose molecules at the non-reducing ends of the oligosaccharides. In the early stages of saccharification α-amylase would produce more of these ends and thus act synergistically with amyloglucosidase. We have just learnt that α-amylase activity is also present in fungal amyloglucosidase preparations (due to acid α-amylase). After initial rapid saccharification, the rate decreases due to low activity on α-1,6 linkages. An enzyme which cleaved these bonds would convert branched dextrins to linear dextrins. These could be rapidly hydrolysed by amyloglucosidase.

6.8.2 Characteristics of pullulanase

pullulanase Pullulanase (EC 3.2.1.41.) is an enzyme which acts specifically on the α-1,6 linkages in branched dextrins. Each of the α-1,6 linked chains must have at least two α-1,4 linked glucose units in it. The smallest molecule on which pullulanase will act is therefore a tetrasaccharide (4-molecules).

∏ Would you class pullulanase as an endo-enzyme or an exo-enzyme? To answer this, make a little drawing of the substrate and think of which bond the pullulanase hydrolyses.

The enzyme acts on the internal α-1,6 bond in branched dextrins. It is therefore an endo-enzyme.

The enzyme is produced from the bacterium *Bacillus acidopullulyticus*. It is stable and active at 60°C and has a pH optimum of 4.5, and so it is ideal for inclusion into the industrial saccharification processes.

The effect on saccharification of replacing half of the amyloglucosidase with pullulanase during an experimental trial is shown in Figure 6.11. This figure also shows the effect of supplementing the amyloglucosidase preparation with acid α-amylase.

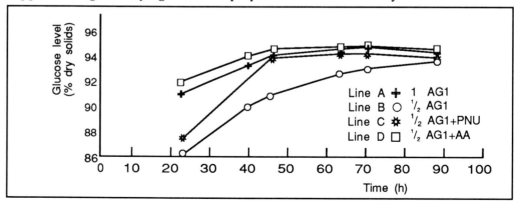

Figure 6.11 Effect of addition of acid α-amylase (AA) or pullulanase (PNU) on saccharification by amyloglucosidase 1 (AG1); pH,4.2, 60°C 33% dry solids.

∏   Use the data in Figure 6.11 to answer the following questions. 1) What effect does
     replacing half of the amyloglucosidase with pullulanase have on a) the total % of
     glucose produced and b) the time it takes to produce maximum glucose levels?
     2) What effect does replacing half of the amyloglucosidase with acid α-amylase
     have on a) the total % of glucose produced and b) the time it takes to produce
     maximum glucose levels? 3) Do you think these differences would have
     significant impact on an industrial saccharification process?

Our responses are:

1)   Comparing lines A and B in Figure 6.11, you can see that halving the
     amyloglucosidase level lowers the rate of saccharification and level of glucose
     reached after 90h. Replacing half of the amyloglucosidase with pullulanase
     (graph C) produces a final glucose concentration similar to that with
     amyloglucosidase on its own (graphs A and B). Maximum glucose levels are
     reached earlier than with the reduced level of amyloglucosidase on its own (about
     70 hours as compared to 90 hours) and at the same time as amyloglucosidase at
     its original level.

     In practice, these results mean that using pullulanase supplement, you can halve
     the amyloglucosidase dose and achieve higher levels of glucose conversion by the
     time maximum glucose levels are reached.

2)   Replacing half the amyloglucosidase with acid α-amylase gives greater levels of
     glucose compared to using amyloglucosidase at its original level (lines A and D
     in Figure 6.11). This means that a higher final level of glucose can be reached, or
     a particular glucose level can be reached in a shorter time.

3)   An increase in attainable glucose levels from 94.6% to 95.2% might not seem
     much. But you should note that the annual world-wide production of glucose is
     5×10^9 (5 billion) kg. An increase in production efficiency of 0.5% thus represents
     25×10^6 (25 million) kg glucose! Whether or not it is economically worthwhile
     depends on how production costs are affected by altering saccharification
     procedures (extra or saved costs due to alterations in enzymes requirement and
     time and labour involved).

The concentration of dry solids in the saccharification mixture does, as we have
mentioned already, affect the final glucose concentration attainable. However, the
system needs to be operated using as concentrated reactants as possible, so that the
volume is minimised: this reduces processing costs (particularly the cost of drying).

Figure 6.12 shows the effect of acid α-amylase addition to amyloglucosidase on the
maximum glucose levels attainable with various substrate concentrations.

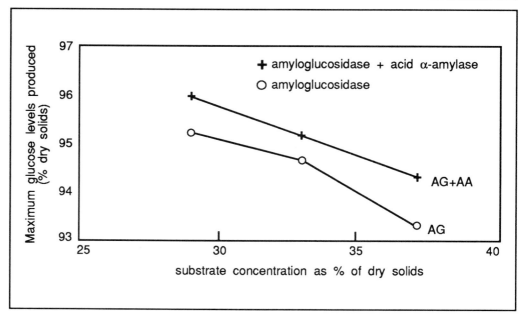

Figure 6.12 The influence of substrate concentration on glucose production during saccharification by amyloglucosidase (AG) and AG with added acid α-amylase (AA).

Π   Use the data in Figure 6.12 to answer the following questions. In industrial saccharification processes, the dextrins are used at an initial concentration of 33% dry solids. When the natural acid α-amylase levels are supplemented (tenfold increase): 1) by how much is the maximum glucose level increased at 33% dry solids; 2) by how much can the concentration of dry solids be increased, while still producing the equivalent maximum glucose level achieved with unsupplemented amyloglucosidase?

The answers to these are:

1)   at 33% dry solids, acid α-amylase supplementation increases maximum attainable glucose levels by about 0.5%;

2)   at 33% solids, amyloglucosidase produces maximum glucose levels of about 94.7%. This glucose level is attainable after supplementation with acid α-amylase at about 36% dry solids.

Pullulanase also enables saccharification to be carried out at higher substrate concentrations.

6.8.3 The saccharification process

The industrial saccharification process is summarised in Figure 6.13. The addition of either pullulanase or acid α-amylase to commercial amyloglucoside has the benefit of:

•   producing higher maximum glucose levels;

•   reducing the amyloglucoside dose required;

- enabling higher dry solids concentration to be used;

- possibly enabling shorter saccharification times.

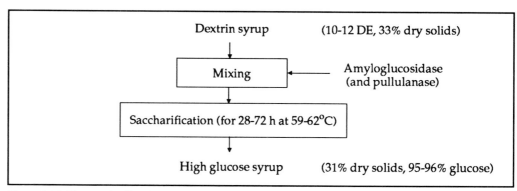

Figure 6.13 The saccharification process.

We have seen that glucose is not as sweet as sucrose (Table 6.1). However because of its greater solubility and other factors, glucose and glucose syrups are used as sweeteners by the food industry, and as a basis for microbial fermentation for products such as food-grade microbial protein. However, it would be advantageous to convert glucose to its isomer fructose, as fructose is sweeter than either glucose or sucrose (Table 6.1). After you have attempted the following SAQs we will examine the applications of enzymes to the isomerisation of glucose to fructose.

SAQ 6.3

Which of the following would produce glucose from amylose most rapidly?

1) α-Amylase (1 dosage).

2) Amyloglucosidase (1 dosage).

3) Pullulanase (1 dosage).

4) α-Amylase (1/2 dosage) and amyloglucosidase (1/2 dosage).

5) Pullulanase (1/2 dosage) and amyloglucosidase (1/2 dosage).

6) Pullulanase (1/2 dosage) and α-amylase (1/2 dosage) and amyloglucosidase (1/2 dosage).

Which of the following would produce glucose from amylopectin most rapidly?

1) α-Amylase (1 dosage).

2) Amyloglucosidase (1 dosage).

3) Pullulanase (1 dosage).

4) α-Amylase (1/2 dosage) and amyloglucosidase (1/2 dosage).

5) Pullulanase (1/2 dosage) and amyloglucosidase (1/2 dosage).

6) Pullulanase (1/2 dosage) and α-amylose (1/2 dosage) and amyloglucosidase (1/2 dosage).

An enzyme, A, when incubated overnight with soluble starch solution at 60°C, produces glucose (as measured in a qualitative test for reducing sugar). Another enzyme, B, shows the same activity.

Enzyme A, incubated overnight with soluble starch solution at 120°C does not produce glucose. A mixture of enzyme A and B does produce some glucose from soluble starch after incubation overnight at 120°C. Identify A and B from the following list.

1) α-Amylase (from *B.licheniformis*)

2) Amyloglucosidase

3) Pullulanase

6.9 The isomerisation of glucose

6.9.1 The characteristics of glucose isomerase

glucose
isomerase

Glucose isomerase (EC 5.3.1.5.) catalyses the conversion of glucose into its isomer fructose (Figure 6.14). Glucose isomerases have been recognised for some time in bacteria such as *Arthobacter spp.*, *Streptomyces spp.*, *Bacillus spp.* and *Actinoplanes spp.* Enzymes initially isolated were of low activity, were not thermostable and required cofactors. However in the 1960s, workers in Japan isolated organisms producing active, thermostable glucose isomerases which were cofactor-independent. This enabled efficient continuous processes using immobilised enzyme to be developed.

The advantages of using enzyme immobilisation can be liseted as:

• reuse of expensive enzyme;

• easier downstream processing;

• operation of continuous systems (which are more efficient than batch systems);

• increased enzyme stability.

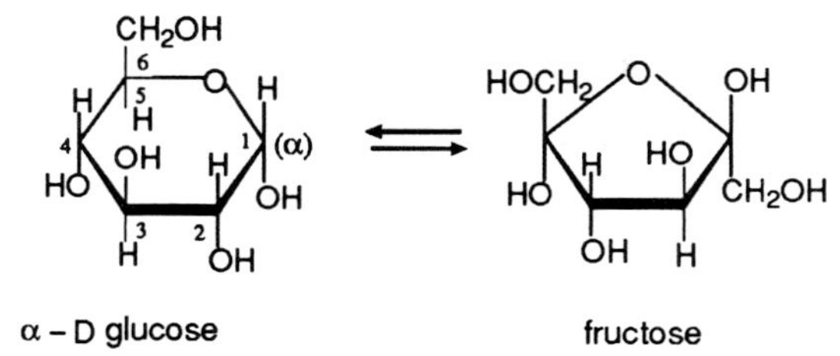

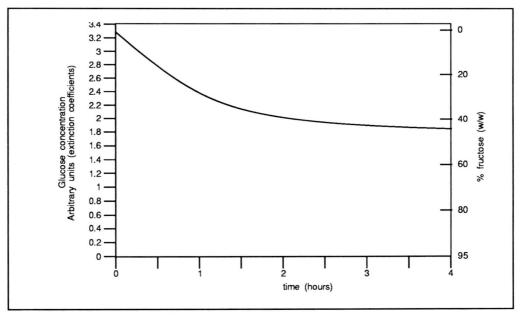

Figure 6.14 The conversion of glucose to fructose catalysed by glucose isomerase.

6.9.2 The immobilised enzyme

The kinetics of fructose formation from a 95 DE high glucose syrup in a batch system are described in Figure 6.15.

Figure 6.15 The isomerisation of glucose syrup (95 DE, 45% dry solids) using immobilised glucose isomerase at 60°C (batch system).

Π How long does the system take to reach dynamic equilibrium, and at what % of fructose?

According to the data available in Figure 6.15, the system reaches dynamic equilibrium between glucose and fructose after about 4 hours, and at about 44% fructose.

compromises on yield   Under normal working conditions, an equilibrium between glucose and fructose of about 51% glucose and 49% fructose is possible. However, as shown in Figure 6.15, the rate of reaction slows considerably as equilibrium is approached, so in practice shorter reaction times are used and a solution of 42% fructose accepted.

Figure 6.16 shows the effect of ageing (for reused immobilised glucose isomerase) on the residence time that a particular system required for conversion to 42% fructose. You will notice that as the system ages, the time for this conversion increases.

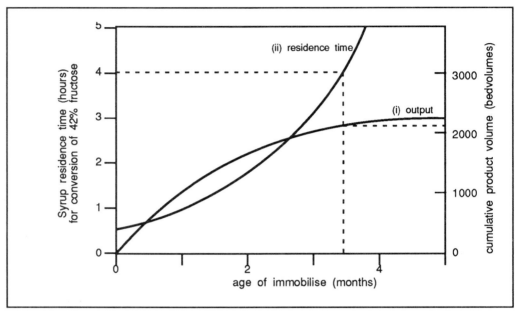

Figure 6.16 The effect of time on i) the output of an immobilised enzyme column and ii) the time required for conversion of 42% fructose (the dotted line relates to the figures referred to in the text).

∏ Why should the required residence time in the enzyme system increase with the age of the immobilised enzyme column?

Enzyme activity in an immobilised enzyme system will always decrease with time, due to factors such as thermal deactivation, oxidation, loss of enzyme, microbial degradation, breakdown of the support material and so on. This means that the residence time to achieve 42% fructose will increase with the age of the immobilised enzyme column.

Because of increased residence times with age, the output (product per unit time) of the system declines (Figure 6.16). For this reason the immobilised enzyme is used only up to a point where optimum (rather than maximum) product has been formed. For the system described in Figure 6.16, this would be after about 3 months or so, when just over 2000 bed volumes, (reactor volumes) of product have been formed, and the residence time has reached 4 hours.

6.9.3 The isomerisation process

typical scale of operation

Typical immobilised glucose isomerase processes operate with the enzyme encapsulated in 1-2 mm particles in a packed-bed column reactor (Figure 6.17). Columns can be 1.5 m in diameter and 4-5 m in height. Such a system would contain up to 4000 kg enzyme, and would produce up to 5×10^5 kg of fructose (dry weight) per day.

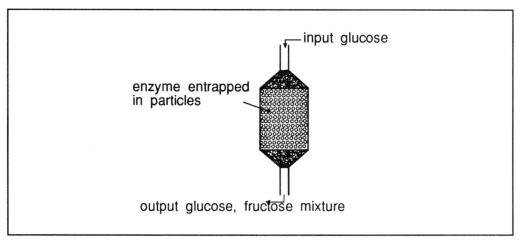

Figure 6.17 Packed bed reactor.

High glucose syrup, HGS (DE 95), which has been concentrated to 45% dry solids and adjusted to pH 7.4 by sodium hydroxide addition, is pumped continuously down the column. The product is called high fructose corn syrup (HFCS). The sugars present in HFCS are fructose 42%, glucose 54% and oligosaccharides 4%.

The flow rate and temperature of the system are determined by several factors. Flow rate must be appropriate to give a sufficient residence time to allow production of 42% fructose. Low feed rate increases the risk of microbial contamination and too-high feed rates cause channelling through the support material. Temperatures below 55°C increase the viscosity of the syrup and also increase the risk of microbial contamination, while high temperatures reduce the stability of the enzyme. In practice, temperatures of about 55°C are used.

6.10 Increasing the fructose content of HFCS

Although the 42% fructose HFCS (termed HFCS 42) is useful in some applications, many users prefer HFCS with a higher level of fructose. To produce these, fructose is removed by chromatographic separation from HFCS 42. The fructose is added to HFCS 42 until the fructose level is 55%, producing HFCS 55.

6.10.1 HFCS 55 production

HFCS 42 is passed through an industrial-scale cation-exchange column, to separate the glucose and the fructose. The resultant fructose (90% fructose:10% glucose) is mixed with HFCS 42 to give HFCS 55. The byproduct of separation is a syrup containing 87% glucose: 5% fructose: 8% oligosaccharides. This mixture is returned to isomerisation column for further isomerisation of the glucose. HFCS 55 is purified to remove contaminants (such as coloured caramels) by passing through a column of activated charcoal, and then concentrated by vacuum drying to 70-86% dry solids.

SAQ 6.6

The flow diagram below represents the production of HFCS 55 from high glucose syrup. From the list given below add inputs, outputs and processes to the diagram to complete it.

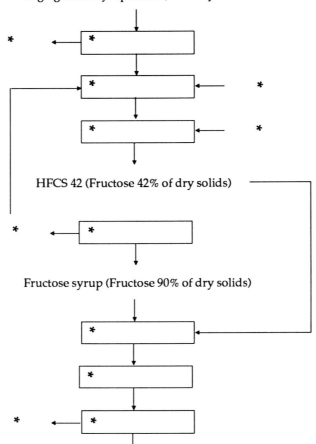

High glucose syrup (DE 95, 31% dry solids)

HFCS 42 (Fructose 42% of dry solids)

Fructose syrup (Fructose 90% of dry solids)

HFCS 55 (at 70-86% dry solids)

Concentration
Glucose syrup (87% Glucose)
Mixing
Chromatographic separation
Concentration (vacuum drying)
NaOH

H₂O
Mixing
Isomerisation
Purification
Immobilised glucose isomerase
H₂O

6.11 The applications of starch-derived sweeteners in the food industry

6.11.1 The physical properties of sweeteners

Let us consider some of the physical properties of the various products we have mentioned and which are of practical importance to food manufacturers.

dextrose

hydroscopic

chewy texture

Glucose syrups can be directly used as food ingredients, or they can be dried to give dry glucose (called dextrose). Glucose syrups are hygroscopic (they absorb and retain moisture). For this reason they (and dextrose) can be used as stabilisers and moisture retention aids. Because of the oligosaccharides present, glucose syrups do have a certain textural component, which contributes to the 'chewy' character of certain confectionery (for example toffee).

reduced A_w

depression of freezing point

High concentrations of glucose and fructose generate low water activities (A_w), and so contribute to the inhibition of microbial spoilage. Glucose depresses the freezing point of ice cream less than sucrose, which is of importance in the production of ice cream and frozen desserts.

Maltose syrup can be produced by the action of fungal α-amylase (EC 3.2.1.1.) on starch or dextrins. Unlike bacterial α-amylase the fungal enzyme acts on small oligosaccharides (DP 5-7), producing syrups with up to 50% maltose. Maltose syrups have a mild sweetness, low viscosity, low hygroscopicity and good thermostability. Maltose sweeteners are popular in Japan.

6.11.2 Applications

Soft drinks

Glucose syrups are very often used in soft drinks in combination with sucrose, because these mixtures seem to have a sweeter taste than each component separately. Dextrose is frequently used as a sweetener in fruit juices and thirst quenchers. Nowadays many carbonated and still beverages in the USA are made with high fructose corn syrups as the sole sweetener. From their introduction, high fructose corn syrups have been used widely in flavoured fountain syrup concentrates (which are diluted with carbonated water and consumed like other carbonated beverages). In still beverages, high fructose corn syrups are often used with other glucose syrups, invert sugar or sucrose. Since the composition of high fructose corn syrups is stable at low pH, the products can be formulated to a sweetness profile that will remain constant in acid foods and drinks. This is in contrast to sucrose, which can suffer hydrolysis (inversion) to invert sugar in acid foods and drinks.

Confectionery

importance of crystallisation

In the confectionery industry it is very important to control the tendency of sucrose to crystallise. Incorporating glucose syrup in the recipe instead of sucrose is a very simple and efficient way to solve this problem as glucose is less prone to crystallisation. At the same time the syrup contributes to the 'mouthfeel' and the body of the confection. Dextrose is used in chewing gum and tabletted confections for its sweetness and also for the cooling effect produced in the mouth because of its negative heat of solution. Anhydrous dextrose is used in chocolate products.

Jams and jellies

flavour
enhancement
lowered A$_w$

In jams and jellies, high fructose corn syrup can replace a major part of the sucrose. The enhancement of fruit flavours is very important, and it seems that fructose does have this effect. The prime function of sugars in jams is, of course, preservation. The high concentration of sugars produces low water activity (A$_w$) which prevents microbial growth. On a weight for weight basis, glucose and fructose are more effective in reducing A$_w$ than sucrose.

Ice cream

minimised
freezing point
depression

Ice cream manufacturers have found some benefits from using high fructose corn syrups along with glucose syrup in ice cream, ice milk and frozen desserts. Freezing point depression is minimised by using high fructose corn syrup together with low DE glucose syrup or by retaining some of the sucrose in the formula. The use of high fructose syrup produces ice cream with a smooth and creamy texture and desirable melt rates. Glucose syrup is also used as a crystallisation inhibitor, and contributes to the 'mouthfeel' of the product.

Canning

greater
succulence

In principle, canned fruit should closely resemble the original fresh product. Glucose syrups were, and nowadays fructose syrups are, used to provide low A$_w$ to help preserve the product, and to provide a 'body' in the cover syrup medium. The syrups also balance the sweetness of the product and penetrate the fruit more effectively than sucrose, making it more succulent.

Baking

humectant
properties

The use of glucose syrups or blends of glucose, maltose and fructose syrups can have a beneficial effect in the manufacture of bread and cakes. Their humectant (water-retaining) properties can help to retain moisture in the products under varying conditions thus preventing staleness or drying out. Protection against moisture loss is particularly important to the keeping qualities of sponge-type cakes. The maltose and dextrose contained in these products is directly fermentable by yeast and therefore dough fermentation can be swiftly and efficiently accomplished. During baking, the glucose syrup can caramelise easily and produce an attractive crust colour on the surface.

Dextrose is also valuable in the preparation of dough baked-goods, such as cakes, rolls, doughnuts, pancakes, biscuits and specialty breads.

Fermentation

wide application

Glucose syrups or high maltose syrups are used as fermentable sugar sources in the fermentation industry. For 'low calorie' beers produced in the USA dextrose is mainly used. Maltose syrups are preferred in the United Kingdom for their contribution to the flavour and the mouth feel of the beers. Glucose syrup can be used as a raw material in fermentation media for the production of processing aids such as enzymes, amino acids, citric acid and xanthan gums (Chapter 3). It is also used in the production of fungal protein destined for use as food. Use in these industries will strongly depend on the desired quality of the end product and the prices of alternative feedstocks.

Pickles and sauces

reduced
process time

High fructose corn syrup is used in sweet pickles and ketchup in high concentrations. In sweet pickles, the rapid penetration of the monosaccharides into the onion or the cucumber helps to reduce the processing time and results in a firm, crisp product. Ketchup made with high fructose syrup maintains a deep red colour.

Meat products

flavour
enhancement

Dextrose is used for its flavour-enhancing properties in a wide variety of meat products, such as sausages, chopped ham, pressed ham, meat loaf, luncheon meats and corned beef.

6.11.3 Consumption of sweeteners

Tables 6.4 and 6.5 give some figures for the quantities of carbohydrate sweeteners consumed. Note the rise in the consumption of high fructose syrups, particularly in the USA. We have used the abbreviations HFS and HFCS. HFS represents high fructose syrup irrespective of the source of the carbohydrate. HFCS represents high fructose syrup derived from corn carbohydrate.

Area	Sweetener	1970	1975	1980	1985
USA	Sucrose	10.4	9.6	9.5	7.4
	HFS	0.1	0.5	2.2	5.1
	Glucose	1.9	2.4	2.4	2.5
EC	Sucrose	10.5	10.0	11.0	11.5
	HFS	-	-	0.3	0.3
	Glucose	0.8	1.1	1.1	1.2
Japan	Sucrose	3.0	2.8	3.0	3.3
	HFS	-	0.1	0.3	0.5
	Glucose	0.8	1.1	1.1	1.2

Table 6.4 Sweetener consumption in the Western World (million tons raw or dry basis).

	Sugar		HFCS	
	1986	1982	1986	1982
Beverage	266	1483	3826	1813
Baking	1432	1186	543	382
Canning	387	209	465	329
Dairy products	447	368	212	168
Confections	1051	849	64	18
Other food and nonfood products	443	604	420	211
Total	4026	4699	5530	2921

Table 6.5 Industrial sugar and HFCS consumption in the USA by market (tons x 1000, dry solids).

Answer true or false to each of the following statements.

1) An important function of HFCS in ice cream is to lower the A_w.

2) Ice cream sweetened with HFCS freezes at a higher temperature than ice cream sweetened with sucrose.

3) An important function of HFCS in pickles is to provide the vegetables with a chewy texture.

4) An important function of HFCS in bakery product is to provide a crust on the product.

6.12 The market for starch-derived sweeteners

The quantities of various sweeteners consumed in the USA, the EC and Japan is shown in Table 6.4. The rise of high fructose syrups has been most striking in the USA. This has been partly due to the adoption of HFCS 55 as a sweetener by the soft drinks industry.

In the EC the situation has been rather different. Here the sugar lobby was stronger than in other farming groups, and to protect sugar producers (farmers), quotas have been established limiting the production of fructose syrup with a fructose content greater than 10%. In a free market, the use of high fructose syrups would undoubtedly be much greater. A similar situation exists in Japan.

The industrial markets for sugar and high fructose corn syrups in the USA is shown in Table 6.5. The projected consumption of sweeteners into the next century is shown in Table 6.6. It is projected that consumption of HFCS will continue to increase at the expense of sugar.

Year	Sugar	HFCS	Glucose	Dextrose	Total Corn Sweeteners	Total
1970	101.7	0.7	14.0	4.6	19.3	121.0
1974	95.6	3.0	17.2	4.9	25.1	120.7
1978	91.4	12.1	17.8	3.8	33.7	125.1
1982	73.7	26.7	18.0	3.5	48.2	121.9
1984	67.6	36.3	18.0	3.5	57.8	125.4
1986	61.0	45.8	18.0	3.5	67.3	128.3
1990**	53.6	47.5	18.0	3.5	69.0	122.6
2000**	40.5	50.5	18.0	3.5	72.0	112.5
2005**	34.7	52.0	18.0	3.5	73.5	108.2

Table 6.6 Projected USA consumption of sweeteners 1970 - 2005 (pounds of sugar equivalent per capita). ** market projections. Note that 1 pound is approximately equal to half a kg.

The situation is complicated by the artificial nature of the sugar market. Only 15% of world sugar production is traded at world market prices, the rest is traded under special agreements (and at inflated prices). It could be that overproduction could lead to the lowering of prices. Recently (1990) the price of corn starch has tended to decrease.

Π   Do you know of a method whereby sucrose (sugar) could be converted into high fructose syrups?

If you do not, re-read Chapter 2. The production of invert sugar is at present uncompetitive with HFCS mainly due to the high substrate cost. If the price of sugar falls in the future, the production of invert sugar could compete with HFCS.

Another complicating factor in predicting market shares for carbohydrate sweeteners is the popularity of low-calorie sweeteners, such as aspartame (Chapter 2). Research and development is already underway to develop even better low-calorie sweeteners.

6.13 Economics of HFCS production

Let us briefly examine the production costs of HFCS. As with products such as beer, the valuable by-products (corn oil, gluten and bran) offset some of the cost of production. We must first deal with some rather confusing units. In the West, the corn industry often deals in bushels. Each bushel weights 56 pounds weight (lb). A pound weight (lb) is roughly equivalent to about 0.5 kg (500g), thus a bushel is about 25 kg. In the corn industry for each bushel of corn processed, approximately 1.5 lbs of corn oil, 14 lbs corn gluten feed and 2.8 lbs of corn gluten meal will be obtained as by-products. Thus 25 kg of corn yields about 0.75 kg of corn oil, 6.3 kg corn gluten feed, 1.2 kg of corn gluten meal. In costing out the production of HFCS 55 we have used kg units.

Table 6.7 shows the production cost of HFCS 55. Prices are in US $^{-1}$ 100 kg^{-1}.

Credit	
by-product credit (value of by-products)	11.00
Debit	
starch cost	8.80
capital depreciation	2.30
labour (2.5 men)	2.00
milling and feed preparation	7.00
refinery utilities	1.00
liquefaction enzyme	0.17
saccharification enzyme	0.35
clarification filter aid	0.12
carbon treatment:	
- dextrose	0.30
- HFCS	0.20
ion exchange:	
- resins	0.23
- chemicals	0.35
isomerisation:	
- enzyme	0.38
- chemicals	0.03
fractionation:	
- resin	0.12
- evaporation	0.16
final evaporation	0.32
Total production cost	**23.85**

Table 6.7 Production cost of HFCS 55. Cost in US $ for production of 100 kg HFCS.

The total cost of 100 kg HFCS 55 is thus US $23.85. Capital depreciation takes account of the cost of the production facilities, and in this instance it is calculated over a period of 20 years (the cost of the facility is spread over this period).

SAQ 6.8

Use the data in Table 6.7 to answer the following. Cross out inappropriate responses.

1) The major production cost in HFCS is the cost of: labour/starch/enzymes/capital depreciation.

2) The value of the by-products from wet-milling reduces the production cost of HFCS by about: 1%/10%/30%/50%.

3) The greatest cost of an enzyme is the cost of: α-amylase/amyloglucosidase/glucose isomerase.

4) The cost of all the enzymes used in HFCS production contributes about: 0.05%/0.5%/5%/50% to the total production cost.

The cost of the corn (starch) substrate varies, and recently has tended to decrease (Table 6.8).

Year	Corn cost (US $ per bushel or 25 kg)	Net starch cost (US $ per bushel or 25 kg)
1981	3.16	1.74
1982	2.48	1.10
1983	3.12	1.62
1984	3.11	1.74
1985	2.52	1.37
1986	1.95	0.79

Table 6.8 The cost of corn and corn starch.

The selling price for HFCS 55 has varied between US $28-40 100 kg^{-1}, depending on the cost of (competing) sugar. The enzymes used worldwide in starch degradation for glucose, fructose and ethanol production were estimated in 1989 to have a value of US $100 million.

6.14 The production of enzymes used in HFCS production

We have seen that HFCS is a bulk low-cost commodity. Thus, the enzymes used in HFCS production must themselves be low-cost. We have also seen that in HFCS production microbial enzymes are used.

∏ Write down a list of at least three reasons why the enzymes used in HFCS production are produced from micro-organisms.

Microbial (as opposed to plant and animal) enzymes are used as they can be produced easily and cheaply by fermentation; they show the required specificity of action. Because of the wide diversity of micro-organisms, enzymes can be isolated which are active at the correct conditions (of pH, temperature and so on). Micro-organisms are also easier to manipulate genetically.

We will not consider the technology of strain selection and enzyme production in detail here. The enzymes used in HFCS production must be food-grade and so must be rigorously tested for toxicity and be produced in food-grade organisms (see Chapter 5).

captive use

The enzymes we have discussed are produced commercially by several companies, including NOVO-Nordisk S/A in Denmark, Royal Gist-brocades NV in The Netherlands and Genencor International in the USA. However, increasingly 'captive use' of enzymes is occurring, whereby the HFCS producers manufacture their own enzymes on site. This is possible due to the straightforward technology involved and the fact that it is not prevented by patents. Using the knowledge you gained from Chapter 5, you could probably draw out a generalised flow diagram for the production of these enzymes.

In general, enzyme production systems involve fed-batch processes using fermenters of 100-200 m^3. Cheap raw materials such as starch, soyabean meal, corn-steep liquor and molasses are used. After initial rapid growth (8-24h), growth is limited by feeding, at slow rates, C or N sources, enzyme production occurs during this phase, for up to 50-200 hours. Glucose isomerase is an intracellular enzyme and so downstream processing involves cell recovery and disruption prior to enzyme purification. The other enzymes we have mentioned are extracellular and so are recovered from the fermentation medium.

6.15 The application of genetic engineering to α-amylase production

increasing
enzyme yield

The enzymes used in HFCS production have, in the main, the required specificity of action and activity, as we discussed. Thus, there is no need to use genetic engineering to insert animal or plant genes into micro-organisms (compare this with chymosin - see Chapter 5). However, genetic engineering has been used to boost the level of production in micro-organisms. α-Amylase is coded for by a chromosomal (and therefore stable) gene in the *Bacillus spp.* we have mentioned. By inserting additional α-amylase genes into such organisms, the level of α-amylase production could be boosted.

plasmid
instability

We will examine the two methods used to produce such organisms with multiple gene copies. One method is to insert the gene into plasmids, and then insert several copies of the plasmids into cells. The disadvantage of this method is that plasmids tend to be unstable. The other method is to insert α-amylase genes into the host-cell chromosome.

6.15.1 Cloning the α-amylase gene from *Bacillus spp.*

The first step towards the goal was the cloning of genes from the chromosomes of several *Bacillus* species encoding α-amylase. A possible strategy is shown in Figure 6.18.

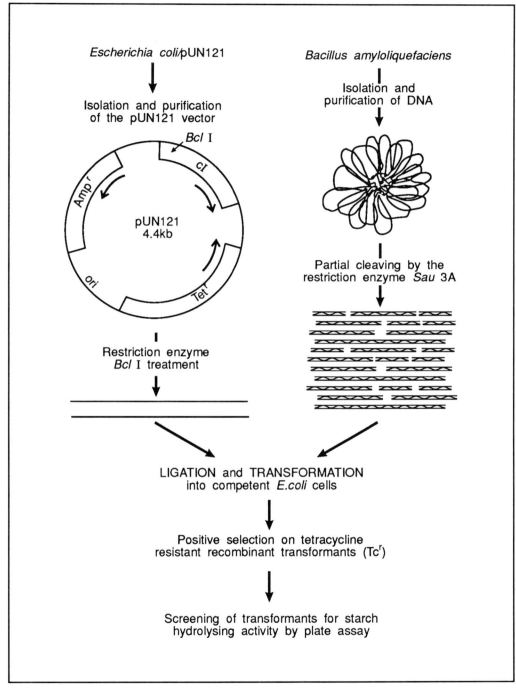

Figure 6.18 Plasmid pUN121 is a positive selection vector for use in *E. coli*. The vector carries an ampicillin resistance gene (Amp^r Tet^s), a tetracycline resistance gene (Tet^r) and a cl repressor gene. Transcription of the Tet^r gene is prevented by the product of the cl repressor gene. Insertion of *Bacillus amyloliquefaciens* DNA into the Bcl I site of the cl repressor gene results in activation of the tetracycline gene. This allows a positive selection of recombinants on ampicillin/tetracycline agar plates. ori = origin.

detection of
α-amylase
producing
recombinants

A cloning vector for *E. coli* is opened with the restriction enzyme *Bcl* I. Chromosomal DNA, cleaved by *Sau*3A restriction enzyme, is ligated into it. After transformation to *E. coli*, colonies harbouring the *Bacillus* α-amylase gene can be recognised by growing organisms on starch agar plates. Such colonies hydrolyse the starch in the surrounding medium. This can be demonstrated by flooding the plate with iodine (Figure 6.19).

Figure 6.19 Colonies grown on 0.4% starch containing agar plates can be tested for the production of α-amylase by colouring with an iodine solution. Clear halos around colonies indicate hydrolysis of starch.

E. coli not
acceptable

The excretion levels of proteins in *E. coli* are low, due to the fact that the organism is Gram-negative and does not easily release such enzymes through the outer layers of the cell wall. Moreover, *E. coli* is not acceptable as a production organism for food-grade enzymes. These problems were overcome by using *Bacillus spp.* as the organisms in which to clone the α-amylase gene.

6.15.2 Transformation of *Bacillus spp.*

Cloning techniques were first developed in *E. coli* a decade ago. Genetic engineering of other bacteria such as *Bacillus*, has proven to be much more difficult. Only after it was found that there were some plasmids such as pUB110 from *Staphylococcus aureus* which could be replicated in *Bacillus subtilis*, that cloning vectors for *Bacillus* could be developed. The first recombinant strains were constructed in *B. subtilis* with a modified pUB110 vector carrying the α-amylase gene from *B. amyloliquefaciens* (see Figure 6.20). Plasmid pKTH10 was cloned into *Bacillus subtilis* strain SB202.

B. subtilis was chosen as a host, since it was the only *Bacillus* species that could, at that time, be transformed with DNA. This host organism has never become a commercial success. More recently, protocols for DNA transformation into *B. amyloliquefaciens* and

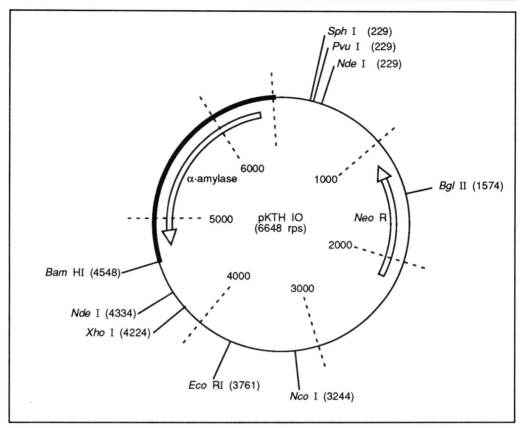

Figure 6.20 Structure of plasmid pKTH 10. This plamsid is a modified pUB110 plasmid.

B. licheniformis have been developed. This allows the use of classic industrial micro-organisms, with good fermentation capacities, as a host for recombinant DNA α-amylase. These strains have a long-standing reputation for safety and are food-grade.

segregational and structural instability   One of the problems encountered when cloning into *Bacillus* is the instability of plasmid vectors. Both plasmid loss (segregational instability) and plasmid deletions (structural instability) occur. This encouraged researchers to develop more stable systems for gene cloning in *Bacillus*. One of these aims at the integration of multiple gene copies into the chromosome. Figure 6.21 describes such a procedure for *B. licheniformis*.

Use Figure 6.21 to follow the description of the process.

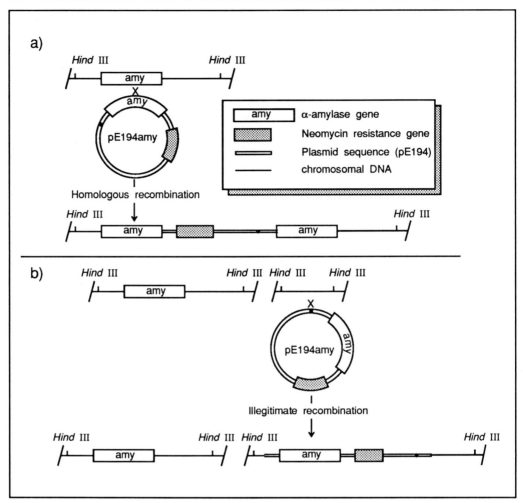

Figure 6.21 Integration of α-amylase genes into the chromosome. Integration into the chromosome is obtained by the use of the thermosensitive origin of replication of plasmid pE194 (see text). There are two possible mechanisms for integration: homologous recombination (Figure 6.21a) and illegitimate recombination (Figure 6.21b). Homologous recombination takes place due to the homology of the cloned α-amylase gene on the plasmid and the resident α-amylase of the chromosome. The result is a tandem duplication of α-amylase genes. Illegitimate recombination takes place irrespective of homology between the α-amylase genes. The plasmid can recombine with any locus on the chromosome. The result is a non-linked array of the two α-amylase genes into the chromosome.

insertion of
α-amylase
gene into the
chromosome

The α-amylase gene is cloned into vector pE194 to produce a new vector, pE194amy. This vector carries a gene for neomycin resistance and has a thermosensitive origin of replication. This origin of replication is active at 37°, but is inactive at 50°C. After transfer of the vector into *B. licheniformis*, the organisms are grown at 50°C in medium containing neomycin. Only bacteria which have one or more linearised copies of the vector inserted into the chromosome will survive this selection. These new strains carry two or more α-amylase genes in the chromosome (the original one plus one or more new copies.). They do not suffer from genetic instability as would recombinants carrying the amylase gene on free (unintegrated) plasmids.

One of the newer α-amylase types (α-amylase, from *Bacillus stearothermophilus*) has only become a success with the aid of genetic engineering. Because enzyme production levels of *B. stearothermophilus* strains are relatively low, commercial production was carried out using genetically engineered *Bacillus subtilis*. This product, was in fact the first enzyme produced by recombinant DNA technology to obtain a GRAS status. GRAS (Generally Recognised As Safe) status can only be obtained after extensive safety testing. Without GRAS status, an enzyme cannot be applied to food processing. Up to now, several GRAS status applications for recombinant *Bacillus* strains producing α-amylase have been accepted by the FDA (Food and Drug Administration, USA). The fact that only natural *Bacillus* DNA is present in these strains is a convincing argument for their safety.

SAQ 6.9

Answer true or false to each of the following statements.

1) The α-amylase gene from *E. coli* has been cloned in *B. subtilis*.

2) The α-amylase gene from *B. amyloliquefaciens* has been cloned in *B. subtilis*.

3) Although the α-amylase gene has been cloned in *B. licheniformis*, production (excretion) levels are low.

4) The gene for thermostable α-amylase from *B. stearothermophilus* has been successfully cloned in *B. subtilis*.

SAQ 6.10

Answer true or false to each of the following statements.

1) Organisms with multiple gene copies of α-amylase can only be produced by integrating the genes into the host cell chromosome.

2) The objective of producing an organism with multiple gene copies of α-amylase is to boost the levels of α-amylase production.

3) The objective of producing an organism with multiple gene copies of α-amylase is to obtain α-amylase from organisms which do not normally produce it.

4) Organisms with multiple gene copies of α-amylase are recombinants.

6.16 Characterisation of α-amylase structure

gene and
amino acid
sequence of
α-amylases

Gene cloning has not only been used to improve production, but has also yielded fundamental knowledge of the properties of α-amylases and on the way they are produced by host cells. The gene sequences of the most important *Bacillus* α-amylases have been determined and from these DNA sequences the amino acid sequences of the proteins have been derived.

identification of
thermostability
and specificity
regions

hybrid enzymes

From these studies, the regions of the enzyme molecule involved in thermostability and specificity of action have been determined. Hybrid enzymes have been produced by recombinant DNA technology. One such patented enzyme is claimed to have the thermostability of α-amylase from *B. licheniformis*, with the preferred hydrolysis pattern of α-amylase from *B. amyloliquefaciens* (which produces less DP 4 and 5 products on which amyloglucosidase has relatively low activity).

These techniques have opened the way to producing more hybrid enzymes and to adjusting the regulatory systems so as to enhance enzyme production and excretion. As the secondary and tertiary structures of the enzymes are unravelled, we may understand more about the basis for substrate specificity and thermostability. This will enable the rational development of improved enzymes using the developing techniques of protein engineering.

SAQ 6.11

Which of the following favour(s) the production of HFCS rather than invert sugar?

1) Starch is cheaper than sugar.

2) Enzymes can be used to manufacture HFCS.

3) Invert sugar contains 42% fructose.

4) Three separate enzymes are required for HFCS production.

SAQ 6.12

The following flow diagram summarizes the production of HFCS from corn starch. From the list below, fit in the processes and associated enzymes at the points marked * to complete the diagram.

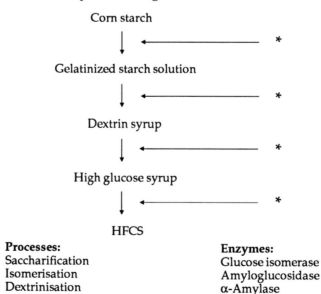

Corn starch

Gelatinized starch solution

Dextrin syrup

High glucose syrup

HFCS

Processes:
Saccharification
Isomerisation
Dextrinisation
Gelatinisation

Enzymes:
Glucose isomerase
Amyloglucosidase
α-Amylase
Pullulanase

Summary and objectives

HFCS represents a good example of biotechnological innovation in food processing. Existing fermentation technology has been combined with genetic engineering to provide a ready and cheap supply of starch-digesting enzymes from micro-organisms. Enzyme immobilisation has enabled the production for the food industry of sweeteners derived from readily available and cheap starch crops. In many respects these sweeteners have better properties than those of sugar.

Now that you have completed the material in this chapter you should be able to:

- describe the uses of sweeteners in the food industry;

- outline the rationale for producing carbohydrate sweeteners from starch;

- describe the chemical structure of starch and its mode of hydrolysis by enzymes;

- describe factors affecting α-amylase activity in dextrin formation;

- describe factors affecting amyloglucosidase activity in glucose syrup production;

- describe the action of pullulanase in glucose syrup production;

- describe the action of glucose isomerase in HFCS production;

- draw and interpret a flow diagram to show the production of HFCS 55 from corn starch;

- describe the application of recombinant DNA technology to improving the production and performance of starch-digesting enzymes;

- outline the economics of HFCS production.

Fruit juices

Fruit juices

7.1 Introduction

Fruit juices have become everyday food items only relatively recently. Initially processing was simply mechanical (such as pressing and filtration). In the 1930s enzymes were applied to juice processing. This use of enzymes was largely empirical (that is enzymes were simply added on a trial and error basis). Since the 1960s the structure of the enzyme substrates and mechanism of enzyme action have been unravelled and enzymes have been used in fruit juice processing in a directed way.

In this chapter we are going to examine the fundamental structure of the cell walls of fruit tissues and the mechanisms by which enzymes act on them. We will also see how this knowledge is applied to improve in the production of fruit juices.

7.2 Composition and structure of plant cell walls in fruits

In order to understand how enzymes affect plant cell walls and influence fruit juice processing, we must first of all review the structure and composition of these walls. We have assumed that you have knowledge of the structure and properties of the carbohydrates which make up the plant cell walls. If you do not, then it would be helpful first of all to brush up on it - you can find this information in the BIOTOL text 'The Fabric of Cells' or in a good biochemistry text book.

7.2.1 Microscopic structure

middle lamella, primary wall, secondary wall

pectin
lignin
cellulose
microfibrils
hemicelluloses

Microscopic investigations have revealed that plant cell walls can be divided into three layers: middle lamella, primary wall and secondary wall (see Figure 7.1). In parenchymatous tissue (which forms the major part of the tissue of fruits) the middle lamella acts as an intercellular binding substance (binding the cells together) and is composed mainly of pectin. Secondary walls contain less pectin but contain some lignin. The primary cell walls consist of cellulose fibres called microfibrils embedded in a matrix of pectins, hemicelluloses and proteins. In woody plant tissues cellulose fibres are usually associated with lignin, but in fruit and vegetable tissues this is absent.

Π What effect do you think the complete removal of pectin from the middle lamella of the cell walls would have on the fruit tissue?

maceration

As pectin is the major component of this binding layer, its complete removal would cause separation of the cells and maceration of the tissue. You must at some time have encountered a vegetable such as a potato or carrot which has gone 'mushy'. This has been brought about by a contaminating micro-organism which has produced pectin-digesting enzymes which have degraded the pectin. This condition is called soft-rot, and is a common cause of vegetable spoilage.

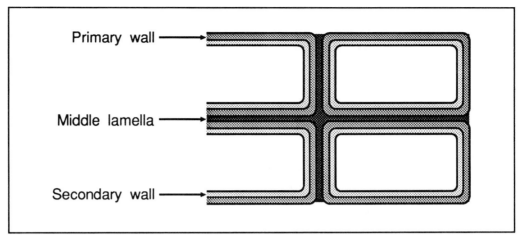

Figure 7.1 Schematic representation of the structure of plant cell walls.

7.2.2 Chemical structure

Pectic substances

pectic substances or pectins

Pectic substances or pectins are a group of closely related polysaccharides, which are deposited mainly in the early stages of the growth of plant cells. Fruit and vegetables are particularly rich in them. Pectins can be extracted from cell walls by treatment with hot water, weak acids or chelating agents. Solutions of soluble pectin have high viscosity, rather like solutions of soluble starch.

molecular components of pectin

The basic structure of pectins is a linear chain of α-1, 4 linked molecules of pyranosyl D-galacturonic acid (Figure 7.2). This linear chain is often referred to as the polygalacturonan (or more simply, galacturonan) backbone. Varying proportions of carboxylic acid groups can be present as methyl esters. The esterification is usually by methanol, (methoxyl pectin), in which case the pectin is said to be methylated. Where more than 50% of the carboxyl groups are methylated the substance is referred to as high methoxyl pectin. Where less than 50% of the carboxyl groups are methylated the substance is referred to as low methoxyl pectin. Where methylation is less than 10% or completely absent, the substance is referred to as pectic acid (or pectate). This knowledge is enough for you to understand the action of the enzymes which we are going to consider. However, you should appreciate that pectin structure is somewhat more complex than has been described above. For instance, the galacturonan backbone also contains a proportion of α-1, 2 linked rhamnopyranosyl molecules and has short side chains composed of up to three molecules of a variety of sugars. The major sugars D-galactose and L-arabinose are present in more complex chains of considerable length with structures similar to arabinans and (arabino)-galactans whose structures we will describe later.

soluble pectin

Some pectin, especially that situated in the middle lamella, can be easily removed (for instance by extraction in water, and is referred to as soluble pectin. This and most of the other pectin in the middle lamella is relatively easily degraded by appropriate enzymes. The pectin within primary and secondary cell walls is not so easily degraded by enzymes.

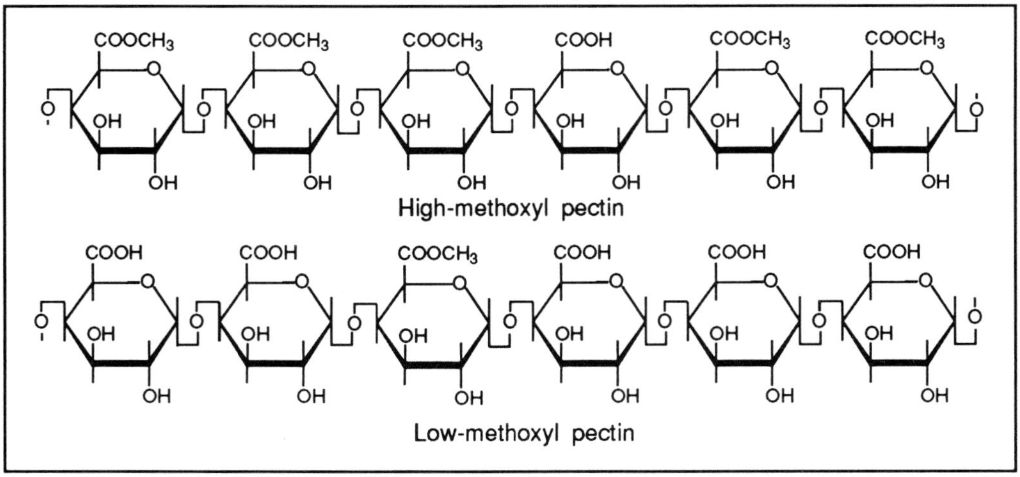

Figure 7.2 The basic structure of pectins.

Hemicelluloses

arabinans
galactans
xylans
xyloglucans

Hemicelluloses are polysaccharides that are extracted from plant cell walls not by weak acids (as are pectins) but by strong alkali. They are composed of four major substances: arabinans, galactans, xyloglucans and xylans.

L-arabino-
furanosyl

Arabinans are branched polysaccharides with backbone of α-1,5-linked L-arabinofuranosyl (Figure 7.3). To about every third arabinose molecule additional L-arabinofuranosyl molecules are attached by α-1,2 or α-1,3 linking. This can produce a structure similar to a comb (with rows of protruding teeth) or a more complex arrangement with multiple branching.

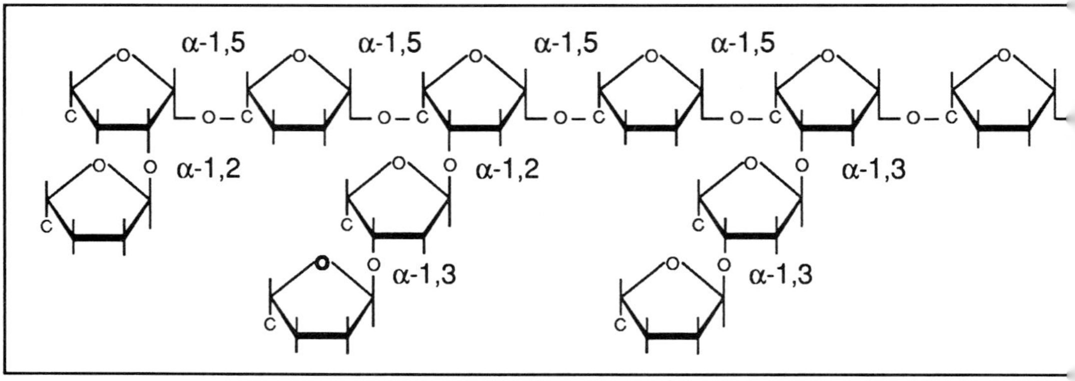

Figure 7.3 The structure of arabinans.

Two groups of galactans can be distinguished: galactans with a backbone of β-1,4 linked D-galactopyranosyl residues and galactans with a backbone of β-1,3 linked D-galactopyranosyl residues substitued at O-6 with β-D-galactopyranosyl residues (Figure 7.4). β-1,3,6 galactans can be substituted at O-3 and O-6 with α-L-arabinofuranosyl groups. β-1,4 galactans can be substituted at O-3 with α-L-arabinofuranosyl groups or with branced L-arabinans. The arabinose containing galactans are called arabinogalactans.

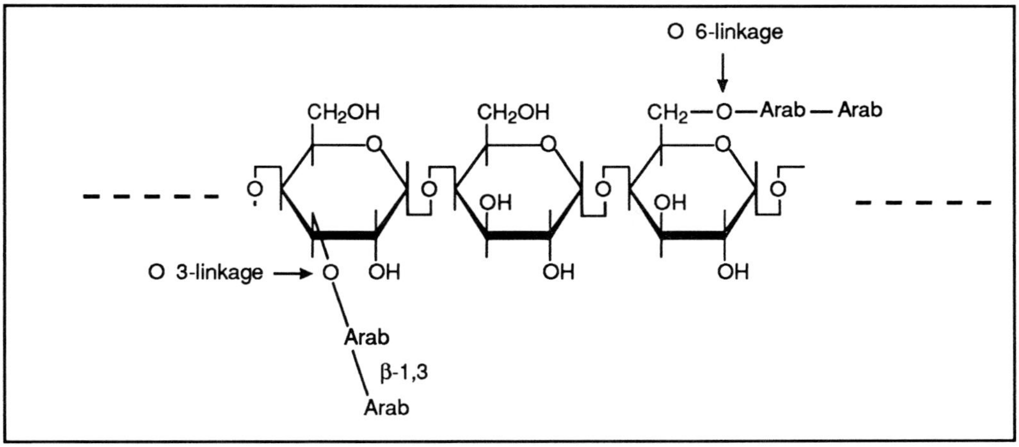

Figure 7.4 The structure of some components of galactans.

Cellulose

cellulose

glucose
residues

microfibrils

crystallinity of
cellulose

amorphous

Cellulose is the best known of all plant cell wall polysaccharides. It is particularly abundant in secondary walls, and accounts for about 20-30% of the dry mass of most primary cell walls. Cellulose is a linear chain of β-1,4 linked glucose moieties (Figure 7.5). In cellulose these β-1,4 glucan chains aggregate by intramolecular hydrogen-bonding along their length to form flat rigid structures called microfibrils. The major structural variable in the celluloses of different tissues is in the degree of polymerisation (the number of units in the chain) of the glucan chains within the microfibrils. A typical value for the degree of polymerisation of cellulose in secondary walls is 14 000. Primary cell wall cellulose appears to occur in two populations, one with a degree of polymerisation below 500 and one with values between 2500 and 4500. The cellulose of secondary cell walls has a rather high degree of crystallinity, with all the glucan chains running in the same direction ('parallel arrangement'). This crystallinity explains the mechanical strength and chemical stability of cellulose. Cellulose in primary cell walls generally has a lower degree of crystallinity. Where the glucan chains are randomly arranged the cellulose is said to be amorphous. This makes the primary cell wall cellulose probably more susceptible to enzymatic degradation.

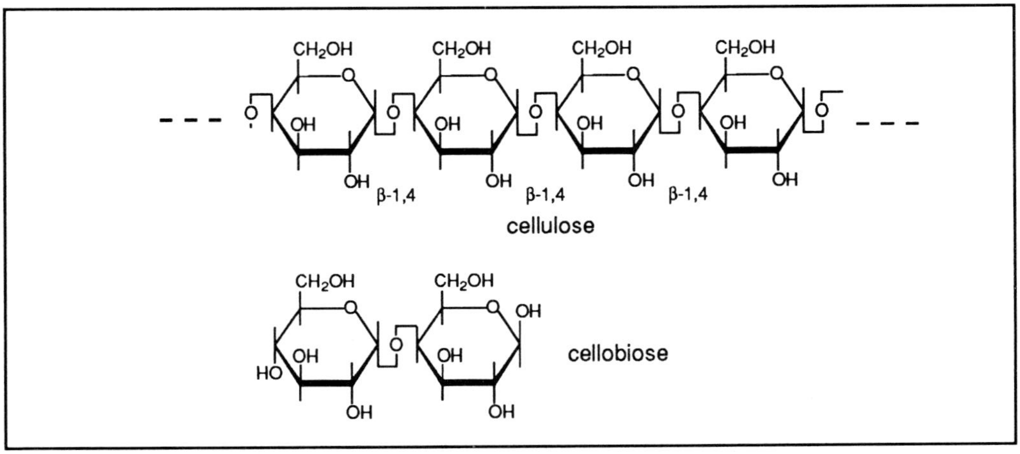

Figure 7.5 The structure of cellulose and cellobiose.

Π   What do you think the effect of degradation of pectin and cellulose would have
    on plant cell walls?

Cellulose forms the rigid structures which give plant cell walls their shape, strength and
rigidity. They are embedded in a matrix of pectic substances, hemicelluloses and
possibly structural proteins. The progressive degradation of cellulose fibrils will lead to
loss of wall strength and eventually its breakdown into fragments. Walls can sometimes
be completely removed of cells resulting in the formation of protoplasts. Use is made of
this in genetic manipulation of plant cells in the processes of protoplast fusion and
electroporation.

7.2.3 A model of the cell wall

There are four major types of polymer in primary cell walls: pectin, hemicellulose,
cellulose and hydroxyproline-rich glycoproteins. Table 7.1 gives you the proportions
of these substances in some fruits.

	% fresh weight of cell (w/w)			
	Pectin	Hemicellulose	Cellulose	Glycoprotein
Cherries	0.51	0.06	0.17	0.32
Pineapple	0.21	0.35	0.27	0.12
Mango	1.02	0.23	0.59	0.32
Apple	0.54	0.34	0.70	0.15
Pear	0.42	0.22	0.40	0.12

Table 7.1 Major components of the cell walls of some important fruits.

7.3 Enzymes which degrade plant cell walls

endogenous
enzymes

Enzymes which degrade plant cell walls can be found in higher plants, and are thus
referred to as endogenous. Many micro-organisms also produce enzymes which can
degrade plant cell wall polysaccharides. They can be used in processing plant material.

Π   Why do you think micro-organisms should have evolved the capacity to produce
    enzymes which degrade plant cell walls?

The most abundant source of nutrients for heterotrophic micro-organisms (those which
need organic carbon) in natural environments is plant material (plant biomass). Thus
the ability to degrade such material offers a significant advantage to micro-organisms.

Π   Would you expect microbial enzymes which are capable of degrading plant cell
    walls to be intracellular (retained within the host cell) or extracellular (released
    from the host cell)?

The plant cell wall is an insoluble and solid substrate. The components of such walls
cannot be taken up by microbial cells and degraded within them. Intracellular enzymes

would thus be of little use. Extracellular enzymes released from the cell would be able to digest plant cell wall components to molecules which can then be taken up and used by the microbial cell.

We will now examine those enzymes which act on the polysaccharides of plant cell walls, and are therefore important in juice processing.

7.3.1 Pectinases

pectin esterases and depolymerases

Pectinases are defined and classified on the basis of their action toward the galacturonan part of the pectin molecules. Two main groups are distinguished, pectin esterases and pectin depolymerases (polygalacturonase, pectin lyase, pectat lyase).

∏ Mark on Figure 7.2 where you would expect 1) pectin esterases and 2) pectin depolymerases to act. Now read on and see if you were correct.

Pectin esterase

mode of action of pectin esterase

Pectin esterase (PE for short) (EC 3.1.1.11) removes methoxyl groupings from methylated pectin (Figure 7.6). This results in the formation of methanol and low methoxyl pectin.

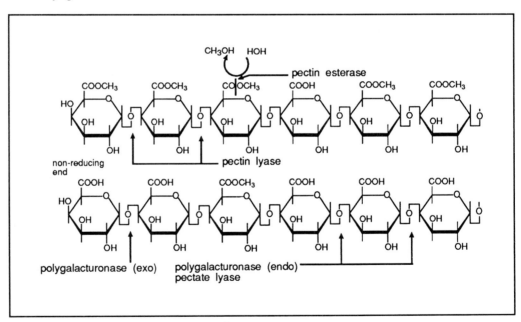

Figure 7.6 The enzymatic degradation of pectin.

Pectin esterase acts preferentially on methoxyl groups of esterified galacturonic acid molecules which are situated either at the non-reducing end of the polygalacturonan backbone chain or (within in the chain) next to a molecule of non-esterified galacturonic acid (that is in which the carboxyl group is free).

Pectin esterases of plant origin proceed along the polygalacturonan backbone in a sequence (like a zip-fastener). This creates regions of non-esterified galacturonic acid. This in itself will not reduce the viscosity of a solution of such pectin as the degree of polymerisation is not affected. In the presence of calcium ions the viscosity may even

Ca$^+$ cross-links and viscosity

increase, as Ca^{2+} cross-links form between long lengths of de-esterified galacturonic acid molecules in adjacent chains. We can represent this cooperative binding of Ca^{2+}, in the following way:

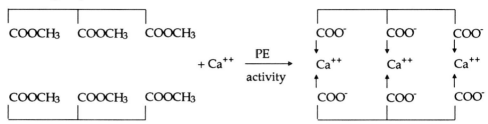

This results in the binding the chains together and so increases the viscosity of the pectin solution. These enzymes have a pH optimum of 7.0 or above, and show stronger affinities for pectins with a lower degree of esterification.

Pectin esterases from fungi remove methoxyl groups from the polygalacturonan backbone in a random fashion (as would occur in chemical de-esterification). These enzymes have pH optima around 4.0–4.5 and have strongest affinity for high methoxyl pectin.

∏ How do you think the activity of pectin esterases from fungi could be measured? (Think of the products of pectin esterase activity).

As the enzyme releases methanol from esterified galacturonic acid and so produces free carboxylic acid groups, enzyme activity can be measured either by appearance of free methanol (measured by gas chromatography or HPLC) or by appearance of carboxyl groups (measured by titration or pH shift). Increased viscosity (measured with a viscometer) on addition of calcium ions to pectin solution can also be used.

Polygalacturonase

poly-
galacturonase

Polygalacturonase (PG for short) exists in two forms: endo-PG (EC 3.2.1.15) and exo-PG (EC 3.2.1.67). Both enzymes are depolymerases and act only on glycosidic linkages between galacturonic acid molecules which are non-esterified. Endo-PG acts on the polygalacturonic backbone randomly, whereas exo-PG only acts on the relevant bond at the non-reducing end of the chain (Figure 7.6).

∏ Would you expect endo-PG or exo-PG to have the greatest effect on the viscosity of a solution of pectin? (Choose one before reading on).

Exo-PG would only sequentially release small fragments from the reducing end of the polysaccharide chains, and so would not greatly reduce the viscosity. Endo-PG would rapidly cause depolymerisation and so reduce the viscosity.

SAQ 7.1

Which of the following substrates would provide the best substrate for endo-PG?

1) high-methoxyl pectin.

2) low-methoxyl pectin.

3) pectic acid.

Both exo-PG and endo-PG occur in higher plants and are produced by micro-organisms. Endo-PG has a pH optimum of around 4.0-4.5 whereas exo-PG has a pH optimum of about 5.0.

Π How do you think the activity of polygalacturonases could be measured?

The activity of endo-PG can be measured by monitoring decrease in the viscocity of a solution of pectic acid or low methoxyl pectin, or by monitoring (by HPLC) the formation of chain fragments (oligomers). The activty of either endo- or exo-PG can be estimated by measuring the increase in reducing end-groups. The two enzymes can be distinguished by using HPLC separation of the products: rapid downward shift of molecular weight distribution (from endo-PG activity) or production of monomers and dimers (exo-PG activity).

Π Can you think of a way in which high methoxyl pectin could be treated in order to make it a suitable substrate for exo-PG?

If you cannot answer this question you must have been day-dreaming when you read the previous section on pectin esterases! The activity of pectin esterase is to make high methoxyl pectin more open to degradation by exo-PG. Such combined activity is an example of synergism.

Pectin lyase

Pectin lyase
specificity

Pectin lyase (PL for short) (EC 4.2.2.10) acts to depolymerise pectins by randomly breaking glycosidic linkages in the polygalaturonan backbone. In contrast to PG, this enzyme acts on glycosidic bonds between galacturonic acid molecules which are esterified (methylated) (Figure 7.6) and it is found only in micro-organisms. The pH optimum is 5.0-6.0. Both the affinity of the enzyme for the substrate and enzyme activity increases as the degree of methylation of pectin increases. As the pH decreases the affinity of the enzyme for low methoxyl pectin increases, and this effect is enhanced by calcium and other cations. This makes pectin lyase a very useful enzyme for fruit processing. The enzyme breaks bonds between methylated galacturonic acid residues producing unsaturated products as shown below:

measurement
of pectin lyase
activity

Activity of pectin lyase can be measured by the decrease in viscosity of solutions of high methoxyl pectin (90% methylation). Alternatively, the degradation products (which have double bonds as a result of the β-eliminative mode of bond breakage) can be measured by absorbance at 235 nm using a spectrophotometer.

Pectate lyase

Pectate lyase

Pectate lyase (PAL for short) (EC 4.2.2.2.) breaks glycosidic linkages between non-methylated galacturonic acid molecules in low methoxyl pectin or pectic acid. The enzyme is found only in micro-organisms. It has an absolute requirement for Ca^{2+} ions and has a pH optimum of about 8.5-9.5 (and has thus very limited activity in most fruit juices). The mode of cleavage of glycosidic bonds is the same as with pectin lyase, and so the same methods can be used to measure enzyme activity, provided low methoxyl pectin or pectic acid is used as substrate.

SAQ 7.2

You are a technician in the laboratory of a factory making fruit juices. Your boss comes in saying "I'm in terrible trouble - I need to know what enzyme this bottle contains but the label has come off and all that is written on the bottle is 'rapidly decreases the viscosity of low-methoxyl pectin solution at pH 4.5 - not active against high-methoxyl pectin'". Write the name of the enzyme on this new label.

SAQ 7.3

Which of the following enzyme mixtures would produce a synergistic effect on the degradation of high-methoxyl pectin? (Use Figure 7.6 to help you answer this)

1) PE and PL.

2) PE and PAL.

3) PL and PAL.

7.3.2 Hemicellulases

Arabinanases

arabinanases
arabinosidase

There are two types of arabinanases: arabinosidase (EC 3.2.1.55) and endo-arabinanase (EC 3.2.1.99). Arabinosidase can be subdivided into two forms, A and B. The B form degrades branched arabinan to a linear chain by splitting off terminal α-1,3-linked arabinofuranosyl side chains (Figure 7.7). At the same time this enzyme slowly and sequentially breaks the α-1,5 links at the non-reducing end of linear arabinan.

endo-
arabinanase

Endo-arabinanase hydrolyses linear arabinan in a random fashion, producing oligomers of shorter lengths. Arabinosidase A degrades the arabinan oligomers to monomers. The enzymes may be prefixed with 'α-L' as they hydrolyse α linkage between L-aribinose residues.

Arabinanases occur in some plants and micro-organisms. Fungal arabinanases have a pH optimum around 4.0 while bacterial arabinanases have a pH optimum around 5.0-6.0.

∏ Would you class arabinosidase B as an exo-enzyme or an endo-enzyme?

The answer is as an exo-enzyme because it releases monomeric arabinose from polymeric and oligomeric arabinans.

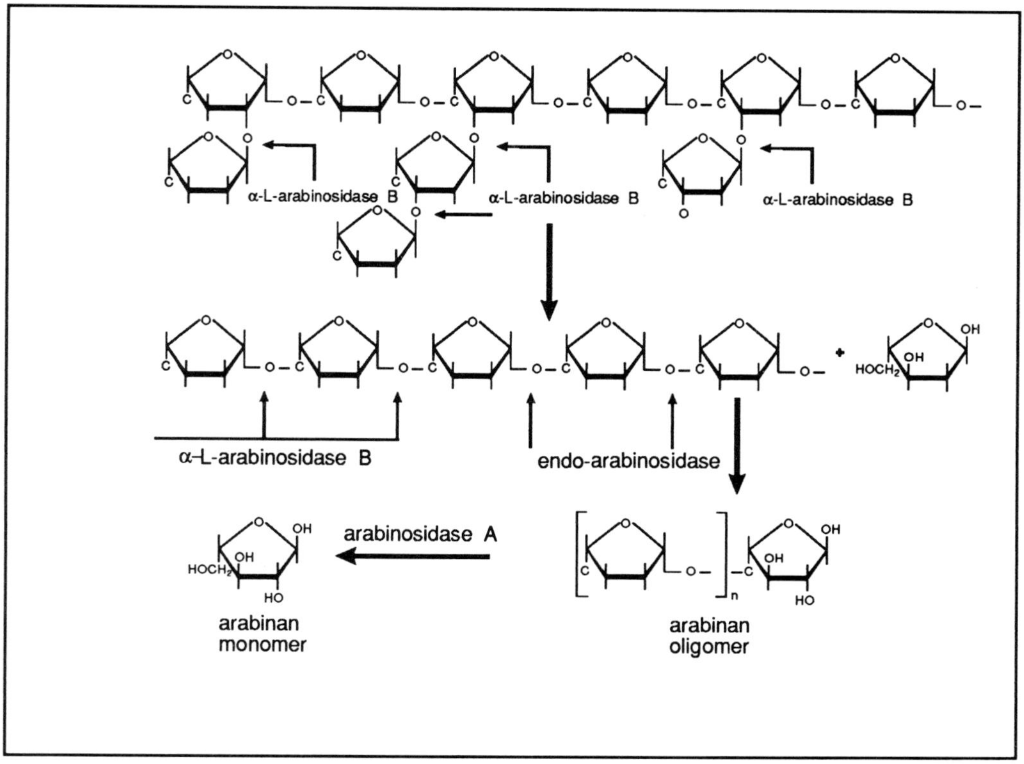

Figure 7.7 The enzymatic degradation of arabinans.

Galactanases

Galactanases Galactanases can also be subdivided into two types: endo-galactanase (EC 3.2.1.89) which randomly breaks β-1,4 linkages of polygalactopyranosyl chains and galactanases (EC3.2.1.90) which break β-1,3 and β-1,6 linkages, also in an endo-fashion (Figure 7.8).

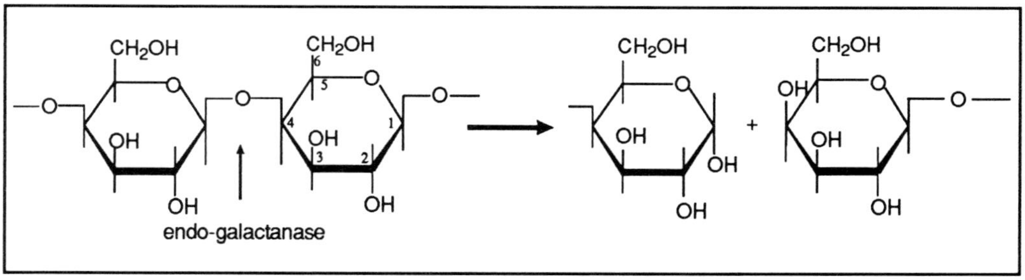

Figure 7.8 The enzymatic degradation galactans.

The activity of the various hemicellulases can be determined spectrophotometrically by the amount of reducing end groups released from their corresponding substrate. The activity of α-L-arabinosidase A and B can be assayed with the appropriate para-nitrophenyl α-arabinofuranoside as substrate. In these assays arabinofuranose linked to para-nitrophenol is incubated with the enzyme. The enzyme hydrolyses this substrate to release arabinose and para-nitrophenol. Free para-nitrophenol is a yellow compound (at alkaline pH) and therefore the release of this moiety by the enzyme can be measured spectrophotometrically. Thus:

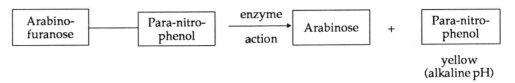

yellow
(alkaline pH)

Other glycosidases can be assayed by using the corresponding para-nitrophenyl derivatives.

7.3.3 Cellulase

cellulase
cellulase
complex

Cellulase (often called the cellulase complex) is a multi-enzyme system composed of several enzymes: endo-glucanase (EC 3.2.1.4) exo-glucanase (EC 3.2.1.91) and cellobiase (alternatively named glucosidase) (EC 3.2.1.21). Endo-glucanase hydrolyses the β-1,4 linked glucan chain of cellulose at random. Exo-glucanase breaks the bonds at the non-reducing end of the chain, producing glucose or cellobiose (dimers of β-1,4 linked glucose - Figures 7.5 and 7.9). Cellobiases split cellobiose into its two component glucose molecules. Enzymes of the cellulase complex are produced only by micro-organisms. Cellobiase is also called β-glucosidase as it hydrolyses β linkages between glucose. Likewise the exo- and endo-glucanase may also be referred to as β enzymes because they hydrolyse β linkages.

Native crystalline cellulose is acted upon by endo-glucanase, which acts primarily on regions of low-crystallinity to form amorphous cellulose. This enzyme and/or exo-glucanase can then easily degrade amorphous cellulose to cellobiose. However cellobiose strongly inhibits the activity of both enzymes.

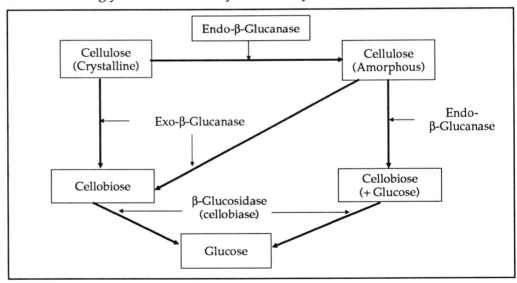

Figure 7.9 The enzymatic degradation of cellulose.

∏ How do you think the inhibition of glucanases by cellobiose during cellulose degradation could be prevented?

The addition of cellobiase would prevent this by removing the cellobiose. This is another example of synergism. However the activity of cellobiase is inhibited by glucose.

The activity of cellulase enzymes can be measured by the production of reducing end groups. The activity is best measured using amorphous cellulose, or carboxymethyl-cellulose (CMC) as a substrate for the enzymes.

SAQ 7.4

An enzyme preparation has the following effects.

It does not produce methanol from high-methoxyl pectin.

It slightly reduces the viscosity of low-methoxyl pectin.

It does not reduce the viscosity of high-methoxyl pectin.

It does not produce reducing sugars from linear arabinan.

It does not produce reducing sugars from carboxymethyl-cellulose.

It produces glucose from cellobiose.

Choose the correct contents of this preparation from the following:

1) PL + exo-glucanase;

2) PG + endo-glucanase;

3) PE + arabinosidase A;

4) cellobiase + arabinasidase B;

5) PG + cellobiase;

6) PL + endo-glucanase.

7.3.4 The enzymatic degradation of cell walls

pulped fruit

Let us examine the effect of adding all the enzymes mentioned so far to crushed fruit (pulp).

The following stages can be distinguished in the enzymic degradation of plant cell walls:

maceration

- Limited degradation of soluble pectin and middle lamella pectin by endo-PG and pectin lyase leading to tissue disintegration and fromation of a (mono-) cell suspension with a certain residual viscosity, the basis of the maceration process;

- Extensive degradation of soluble pectin, middle lamella pectin and cell wall pectin by endo-PG and pectin esterase and/or pectin lyase renders the gelified juice to a thin liquid, however part of the cell walls remain which serve as a pressing aid and facilitate release of the juice, the basis of the pulp enzyming process and juice clarification;

liquefaction

- Total solubilisation of virtually all cell wall polysaccharides by a broad spectrum of pectic, cellulolytic and hemicellulolytic enzymes, the basis of the liquefaction process. The ultimate stage would be total saccharification of the solubilised polysaccharides to their building sugars by oligomerases and glycosidase.

SAQ 7.5

The figure below shows the release of sugars from the isolated cell wall fragments of apple cells after treatment with various enzymes. Soluble pectin has already been removed from these by treatment with a chelating agent and hot water extraction.

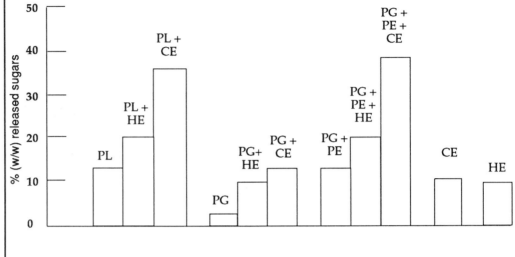

PL	pectin lyase
PG	polygacturonase
PE	pectin esterase
CE	cellulase
HE	hemicellulase

The activites of PL, PG, cellulases (CE) and hemicellulases (HE) alone and in various combinations are shown. Use this data to answer the following. Cross out incorrect options.

1) Significant amounts of pectin are released from wall fragments by: PL/PG/PG+PE.

2) Cellulases are more/less effective than hemicellulases when each is used in isolation.

3) Cellulases are more/less effective than hemicellulases when each is used in combination with a pectinase.

4) The effect of cellulase or hemicellulase is least when it is added to: PL/PG/PG+PE.

5) The maximum effect occurs with the addition of cellulase/hemicellulase to: PL/PG/PG+PE.

6) The effect of adding cellulase to PL or PG + PE is/is not synergistic.

7) The effect of adding hemicellulase to PL or PG + PE is/is not synergistic.

The action of enzymes in degrading cell walls is influenced particularly by temperature, pH and the presence of inhibitors. In addition the physical structure of the wall such as fragment size, degree of crystallinity of cellulose and degree of enmeshment of the various constituents, will influence the accessibility of the substrates to the enzymes. The chemical structure of the walls (degree of methylation, degree of branching, composition of side chains) will also strongly influence degradation. These physical and chemical properties of fruits (and vegetables) vary from species to species and with the degree of ripeness or age.

accessibility of substrates

7.3.5 Sources of commercial enzyme preparations

fungal sources

Pectinases used in fruit and vegetable processing have an annual market value of about US $ 16 million. They are fungal enzymes produced from food-grade species of *Aspergillus*. Traditionally the enzyme preparations used have been rather crude, and have contained pectinase, hemicellulase and cellulase activities. The enzymes have been produced by submerged fermentation in liquid media. This technology is essentially the same as that used for the production of other enzymes (for example recombinant chymosin in Chapter 5). A flow diagram of the process is shown in Figure 7.10. In some processes enzymes have been produced in solid-substrate (or solid-state) fermentation, in which the fungus is grown on solid material such as crushed wheat or oats.

More recently, brands of enzymes with more specific activities have become available. For example purified PG preparations for maceration; PE preparations for clarification; cellulases for liquefaction and hemicellulases for clarification.

food-grade sources

Trichoderma

Pectinases and cellulases are produced by a number of micro-organisms. Only enzymes from food-grade organisms are useful in the content of juice production. The production of such enzymes is most effective by the fungus *Trichoderma*. While such enzymes can be used in Europe they do not (yet) have the approval of the Food and Drug Administration (FDA) in the USA and so are not used here.

We will now examine how fruit juices are produced and how enzymes are applied to the process.

7.4 Fruit juice production

7.4.1 Fruit juices

fruit juices

cloudy

Fruit juices are defined as the liquid expressed by pressure or other mechanical means from the edible portion of fruit. Frequently they are turbid or cloudy, containing insoluble cellular components in colloidal suspension with variable amounts of finely divided tissue. The solid matter content of fruit juices is generally 5-20% (fresh weight). Some juices are consumed in a natural cloudy state and may also contain oily or waxy material and carotenoid pigments derived from the skin or rind of the fruit. Clarification of such juices would impair their appearance and flavour because these are associated with the insoluble cellular material (eg orange juice).

Some fruits are not suitable for direct processing into juice because their flavour or colour pigments are bound mainly to the pulp particles which are sometimes too thick to be drinkable. They are instead often converted into pulp products by disintegrating the whole fruit to give purees, (eg apricot, mango and tomato). Other fruits are unsuitable because they are too acidic (black currant and sour cherry), or have such a strong flavour (passion fruit) that they are unsuitable for direct consumption. From such

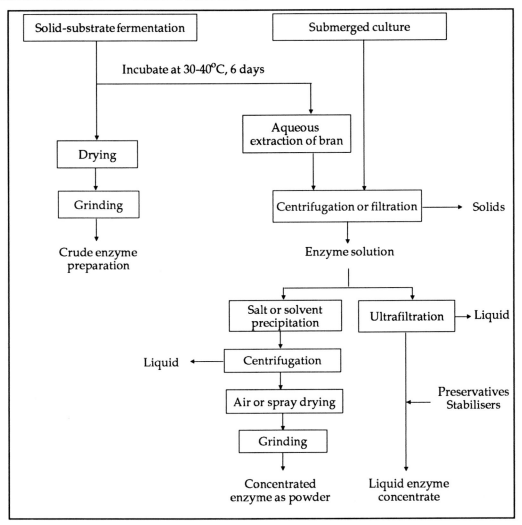

Figure 7.10 The production of fungal pectinases.

fruits, clear or cloudy juices are prepared by adding water, sugar or acid. These juices are called nectars and their fruit portions vary between 25-50% (fresh weight).

clear condition

The juices of most deciduous fruits (apples, berries) and grapes have traditionally been consumed in clear condition. The problem has been to clarify them effectively and to maintain them in that condition throughout their storage life. For cloudy juices and nectars, the stability of the cloud is also an important factor.

7.4.2 Fruit composition

importance of fruit constituents

Fruit composition is influenced by genetic make-up, fruit maturity, cultural and nutritional conditions, climate and weather and post-harvest changes. The suitability of a particular species or cultivar for juice is particular dependent upon the balance of acid and sugar, the type and amount of phenolic components, the aromatic constituents, and sometimes vitamins. All cells contain a vacuole in which water-soluble constituents and compounds are dissolved. The vacuole is enclosed by a semi-permeable membrane composed of lipoproteins. In this way osmotic pressure is created which presses the membrane against the cell walls. The cell walls are rigid but are fully permeable. Upon mechanical disruption, cell walls are fragmented and the vacuoles burst open. A mash is obtained from which the cell sap (raw juice) can be separated by gravity and/or by pressure. Many chemical, enzymatic and physical changes occur as soon as the cell is disrupted.

7.4.3 The traditional process

pulping

raw juice

The traditional way of making fruit juices mainly involves disruption of the fruit by crushing and/or grinding. This transforms the intact organised tissues into a semi-fluid mixture of cells and cell wall fragments in cell liquid. This mixture is called pulp and the process of forming it is called pulping. The raw juice is then separated from the pulp by use of presses or sieves. In more-modern processes, centrifuges may be used. A variety of presses have been developed, including cloth and rack presses, screw presses, belt presses and pneumatic and hydraulic tank presses. The important characteristics of presses are capacity, juice yield (quantity of juice produced from a given quantity of pulp), the amount of sediment in the juice, whether or not the addition of agents (called pressing aids) are required for effective pressing and whether or not they are suitable for continuous operation. Juices produced in this manner will tend to be cloudy.

pressing aids

problems with pectin and viscosity

Clear juices are obtained by additional processing. The quantity and fineness of the pulp can be adjusted by altering the crushing/grinding process. Course fragments are retained during pressing, or can be removed by filtration. However, this is often not effective in producing clear juices. Another problem generally is that some raw juices are viscous, due to the presence of soluble pectin. This means that such juice is difficult to separate from the pulp (the juice yields are low). This is the case with soft fruits like bananas, currants, berries and cherries. Pectin in raw juices might not produce high viscosity in the raw juice, but can lead to viscous (and unmanageable) solutions if the juice is concentrated. Concentration of juices is common practice, in order to make storage, transportation and handling cheaper and easier. Viscosity problems in fruit pulping and pressing can be overcome by heating, but this can adversely affect the quality of the juice.

SAQ 7.6

The viscosity problem which can arise during pulping and pressing of juices could best be reduced by the addition of which of the following?

1) PAL.

2) PG.

3) PE + PG.

4) PL.

5) PL + cellulase.

6) PL + hemicellulase.

7.4.4 The stability of fruit juices

microbial
spoilage

enzymic
mediated
changes

All types of juices are inherently unstable: they rapidly undergo microbiological degradation by organisms already present in the fruit or gaining access to the product during processing. They are also subject to enzymatic and non-enzymic changes. It is essential to destroy the micro-organisms at an early stage, or to prevent their growth by heat treatments or by refrigeration.

∏ Write down a list of the likely changes that micro-organisms might produce in fruit juices (now check your list with our ideas).

Micro-organisms can ferment the sugars in fruit juices to make wines and cider. Yeast will be present in fruit juices, and will ferment them if steps are not taken to prevent it. Lactic acid bacteria are also likely to be present, and could ferment the sugars present to lactic acid. In addition to these activities many other microbial products could be formed, altering the organoleptic properties of the juices. It is likely that cell wall-digesting enzymes could be produced as well.

advantages of
producing
concentrates

Heat treatments or refrigeration of juices also affect chemical and enzymatic changes and restrict the growth of micro-organisms. The storage of juice involves the maintenance of large stocks of finished product in a stable condition in containers for distribution, or the installation of expensive refrigerated tankage. As a result, increasing volumes of juice are now converted into concentrates that are both more stable and require less storage space. The concentrates may be diluted for further processing or they may be distributed unchanged through the frozen food chain. There exists a wide range of juice products. These include:

• chilled juices with a short storage life;

• pasteurized single-strength juices in cans, brick packs, glass bottles and in synthetic polyethylene pouches for direct consumption;

• frozen concentrates for dilution.

Both single strength and concentrated juices are distributed in large containers as feedstock for the manufacture of a variety of soft drinks. Dehydrated juices in powder form are also on the market and are gaining increased interest.

7.5 Enzymes in juice processing

7.5.1 Enzymes and juice clarification

cloud particles

We have described clarification as the reduction in viscosity and removal of cloudy material from extracted fruit juices. It is mostly applied to the production of apple juice, and involves the use of fungal pectinases. PE and also endo-PG are essential activities to degrade high methoxyl pectins in apple juice. Because of the highly esterified pectins, juice clarification can also be obtained by PL.

As well as causing degradation of pectin in solution, the pectinases act on cloud particles. These are positively charged protein complexes which are coated by negatively charged pectin, (Figure 7.11). Pectinases remove some of the pectin coating and so expose the positively charged nuclei. This reduces electrostatic repulsion between the cloud particles and causes them to coagulate and precipitate out of solution. Flocculating agents such as gelatin or bentonite (a type of clay) can be added to enhance the process (this is often done to clarify beer and wines). Ultrafiltration has also recently been introduced as a final step in apple juice clarification. This process also removes contaminating micro-organisms.

gelatin and bentonite additions

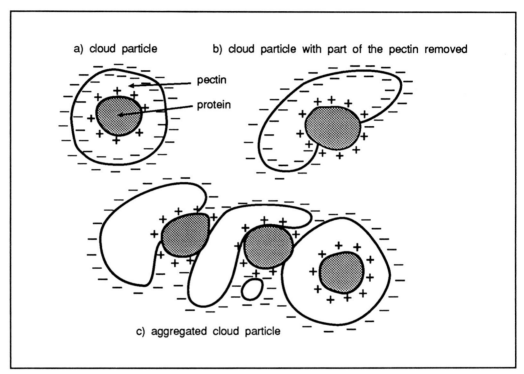

Figure 7.11 The structure of cloud particles. a) intact cloud particle, b) cloud particle with part of the pectin removed, c) positively charged protein causing aggregation of cloud particles.

use of elevated temperatures

Enzymes are added to raw juice at raised temperatures of 45-50°C to reduce reaction times and/or quantity of enzyme required. It also prevents microbial contamination and fermentation of the juice while the clarification is being carried out. Once clarified, the juice can be concentrated without pectin gels being formed. However juice from early-season apples may contain starch, which may thicken after the juice has been heated and/or flocculate during juice concentration.

∏ How do you think the undesirable effect of starch could be overcome?

In Chapter 2 we mentioned amylases, which were used to degrade starch in the manufacture of beer and in Chapter 6 in the manufacture of high fructose corn syrups. Such enzymes, in particular amyloglucosidase, are added to juices for the degradation of unwanted starch. A heat treatment of the juice to gelatinize the starch present is necessary prior to the addition of the enzymes. Enzymes active at 50°C are available, and so can be added with the pectinase.

must

A variation of this enzymatic clarification process has been introduced in the French farm-based cider industry. Traditionally clarification of the crushed fruit (must) has been a spontaneous process, dependent upon the activity of endogenous pectin esterases. These enzymes cause the partial demethylation of pectin which can be flocculated on the addition of Ca^{2+} ions. The flocculated pectins rise to the top of the must (called 'the chapeau brun' - literally the brown cap). This can then be easily separated off. However the activity of endogenous PE varies widely with apple variety and growth environment. Recently fungal PE has been added to enhance the process.

SAQ 7.7

Choose the correct completion to the statement. In the process to produce the 'chapeau brun', fungal esterase preparations which are free of PG or PL activity must be used because:

1) PG or PL activity would break down the pectin to monomers acid, which would inhibit PE activity;

2) PG or PL activity would break down the pectin backbone to oligomers, on which PE is less active;

3) PG or PL would break down the demethylated regions of pectin which PE produces. The products of PG or PL activity would be too small to flocculate with Ca^{2+} ions.

In practice, conditions for clarification are 1-2h at 54°C with an enzyme concentration of about 50 mg l^{-1} (ie 50 parts per million - ppm) or 6-8h at 16°C for the same enzyme concentration. Activity is measured by viscosity reduction or (if PE is included) methanol formation. An iodine test is used to check for starch degradation.

7.5.2 Enzymes and fruit maceration

disadvantages of conventional maceration

Macerated fruit and vegetable preparations (purees) are used in the production of nectars, as ingredients in dairy products and as baby foods (babies do not have the teeth with which to do this for themselves). This process has traditionally been carried out by some thermo-mechanical process (involving mechanical crushing/grinding and heating). However these processes can produce browning in the product (from caramelised sugars produced by the heating - the Maillard reactions), and cause loss of colours and vitamins and production of unwanted flavours (called off-flavours).

blanching

Enzymatic maceration is directed at yielding a product of smooth consistency, but retaining cell walls intact. Endo-PG and PL activities, in the absence of hemicellulase or cellulase activities, are required for this. These cause the separation of cell walls but leave the walls intact. In some circumstances, significant endogenous PG and PE activity can be present in the fruit tissues and this can lead to uncontrolled reactions and some wall degradation. In these circumstances the material can be heated for a short period in order to inactivate endogenous enzymes (this treatment is called blanching).

7.5.3 Enzymes and juice extraction

fruits difficult to press

The pulps of most soft fruits, such as black currants, strawberries, raspberries, bananas and others are by no means easy to press. Treatment of these pulps with pectinases to facilitate pressing and to ensure high yields of juice and pigments has been common practice for the past few decades. These fruits have high pectin contents and give very viscous juices which adhere to the pulp particles as a semi-gelled mass upon mechanical pulping, and from which it is very difficult to extract the juice. When pectinase is added, the gel structure collapses, the juice viscosity decreases and juice can be obtained easily and in high yields. In addition, the structure of the tissue, including skin tissue, is degraded and this allows the colour pigments, which are often enclosed in the skins of berries, to diffuse into the juice so that also the colour yield is greatly improved. The enzyme treatment is often combined with a heat treatment. Enzymatic juice extraction has also become an established process in the production of grape juice and apple juice. It promotes high juice and colour yields from grapes and high juice yields from apples (which are often difficult to press because of variety or prolonged storage). Enzyme treatment of olives, palm fruit and coconut pulp to increase oil yield has also been described.

fruit yield

colour yield

The amount of enzyme required for extraction varies directly with the content of pectin in the fruit. Moreover in fruit such as berries, the high acid content (and so low pH) requires enzymes with stable activity at low pH. The enzymes required for extraction are the same as those required for clarification. It is thought that they act by degrading pectin and cause it to loose its ability to bind water, which therefore appears as juice. In practice conditions for extraction are 30-60 minutes at 15-30°C with an enzyme concentration of 50-200 mg enzyme l^{-1}. Activity is measured by monitoring the rate of juice release, juice yield, by methanol formation (if PE is included) or by filtration tests which measure the amount of sediment in the juice.

7.5.4 Enzymes and juice liquefaction

Through the use of pectic, hemicellulolytic and cellulolytic enzymes the cell walls of fruit pulps can be degraded to the stage of almost complete liquefaction. This has led to the introduction of the enzymic liquefaction process for (tropical) fruit and vegetable pulps. Juice yields exceeding 90% can be obtained and, as a result of the solubilisation of cell wall polysaccharides, the dry matter content of the juices is also usefully increased. The methods used for liquefying juice or pulps from tropical fruits vary widely according to their structure and composition. The fruits themselves are often of inconvenient shape or size. Depending on the accessibility of the cell wall components to the enzymes almost clear (papaya), cloudy (peaches) or pulpy (carrots) juices are obtained. After enzyme treatment, juices can be separated by centrifuges or simple filters (so avoiding the use of expensive fruit presses). Raw juices can be clarified by the methods outlined in the previous section.

accessibility and nature of product

treatment of pomace

During the pressing of pulped apples for the production of apple juice, the juice is extracted by pressing. The material remaining is referred to as pomace. This can be treated by the liquefaction process in order to extract additional juice. Liquefaction is best achieved by the use of pectinases in combination with both endo- and exo-1,4-glucanases. However some quality changes have been observed for juices obtained by the liquefaction process:

- changes in flavour due to the release of flavour precursors and flavour-releasing enzymes within the cell walls (and which otherwise would be retained in the press residue);

- increased liability to non-enzymatic browning when exposed to excessive heat treatment or inadequate cool storage (unsaturated oligomeric pectin fragments released by lyase activity are thought to be responsible for this);

- increased acid content;

- increased susceptibility for haze formation (see below).

An attractive feature of the use of enzymes is its independence of the size of operation: it is as suitable for small operations as for very large ones.

7.5.5 Enzymes and haze reduction

haze

A problem which often occurs upon the concentration of the juice of apples or pears which has been produced by intensive enzyme mash treatments (such as liquefaction), or high temperature processing, is that of haze formation. These processing treatments result in increased quantities of arabinans in juices. These arabinans are normally soluble and do not form haze. However during heat treatment, if appropriate enzymes are present, the side chains from the branched arabinans can be removed. The resultant linear arabinan chains can associate in concentrated juices and form insoluble crystals (ie haze).

SAQ 7.8

Write 'cause' against whichever of the following enzyme/enzymes mixtures would cause haze formation after enzyme mash treatment, and 'prevent' against that whichever would prevent it.

1) endo-arabinanase.

2) arabinosidase A.

3) arabinosidase B.

4) endo-arabinase + arabinosidase A.

5) endo-arabinase + arabinosidase B.

7.5.6 Endogenous enzymes and juice processing

We have already learnt that endogenous pectinase enzymes exist in plants and that these can cause unwanted effects during maceration. Endogenous enzymes are also important in a number of other ways.

effects of endogenous pectin esterase texture and firmness

Endogenous pectin esterase in citrus fruit can cause serious quality defects such as cloud loss on standing of citrus juices, or the formation of a calcium pectate gel in the case of concentrates. Proper processing of concentrated juice including pectin esterase inactivation by heat treatment (avoiding damage of the heat-sensitive flavour), and frozen storage and transport overcomes these problems. Where this sophisticated technology is not available, addition of endo-polygalacturonase to the system can extend the cloud stability of these products and also facilitate making concentrates. Endogenous pectin esterase can also be exploited for producing and improving the texture and firmness of several processed fruits and vegetables (apple slices, canned tomatoes, cauliflower, carrots, potatoes, beans and peas). Blanching at optimal temperatures allows pectin esterase activity to partially de-esterify the cell wall pectins,

tomato processing

which then react with (added) calcium ions resulting in stronger intercellular cohesion. In tomatoes, pectin esterase and endo-polygalacturonase are abundantly present. The presence of these enzymes has a great impact on tomato processing. For the manufacture of tomato products of which a high viscosity (juice) or high consistency (paste) is desirable that the endogenous enzymes be inactivated as quickly as possible. This is achieved in special equipment in which the tomatoes are crushed directly in circulating hot (90°C) tomato pulp. Such juices are characterized as 'hot break'. For applications of tomato products where the consistency is not important, or provided by other ingredients, the tomatoes are crushed and held between 40 and 60°C to ensure breakdown of the pectic material by combined pectin esterase/polygalacturonase action prior to further heat processing. The resulting thin liquid 'cold break' juice can easily be concentrated for making tomato sauce and tomato juice.

7.6 Future trends

improved
enzyme
specification
and yield

Our increased knowledge of the enzyme-based processes has created a need for better formulated enzymes containing specific, technological relevant enzymes, devoid of undesirable activities. Novel purification techniques in the removal of undesirable activities are being developed in the downstream processing of crude enzyme preparations. Another approach is to boost the production of the desired enzymes by recombinant DNA technology (genetic manipulation) and to improve their properties (pH optimum, temperature stability) by site-directed mutagenesis and express the (modified) genes in already tailored industrial (food-grade) micro-organisms. It can be expected that the availability of a whole range of specific enzymes will also lead to new applications for such enzymes.

Summary and objectives

Microbial enzymes are used in various stages in the processing of fruit and vegetable juices. Pectinases are mainly applied and are used to improve the juice yields and characteristics of juices, and in the production of purees. Although currently-used enzyme preparations are fairly crude mixtures of enzymes, purification procedures and genetic engineering are being applied to the production of pure enzyme preparations.

Now that you have completed the material in material in this chapter you should be able to:

- describe the chemistry and arrangement of pectin, hemicellulose and cellulose in plant cell walls;

- describe the modes of action of relevant enzymes involved in the degradation of pectin, hemicellulose and cellulose;

- explain how each of the relevant enzymes can be usefully applied to fruit juice production;

- explain the synergistic activity of relevant enzymes;

- explain how synergistic enzyme activity can be usefully applied to fruit juice production;

- outline the sources of relevant enzymes.

Amino acids: their production and uses in the food industry

Amino acids: their production and uses in the food industry

8.1 Introduction

Amino acids have obviously always been important in foods, as essential nutrients. However their use as food additives also has a long history. For centuries in Japan an extract of seaweed called konbu has been used as a flavour enhancer, to improve the taste of foods to which it was added. Today the active ingredient, monosodium glutamate, is produced in large quantities by microbial fermentation processes. As well as fermentation processes, enzymatic synthesis and conversion is increasingly employed in amino acid production. In this chapter we will examine the developments which have occurred in amino acid production and the uses to which amino acids are put in the food industry.

8.2 The amino acids

If you are unfamiliar with the amino acids and the common notation for them, refresh your memory by examining Figure 8.1. We do not expect you to instantly remember all of these, but as you read this chapter you might find this figure a useful reference source.

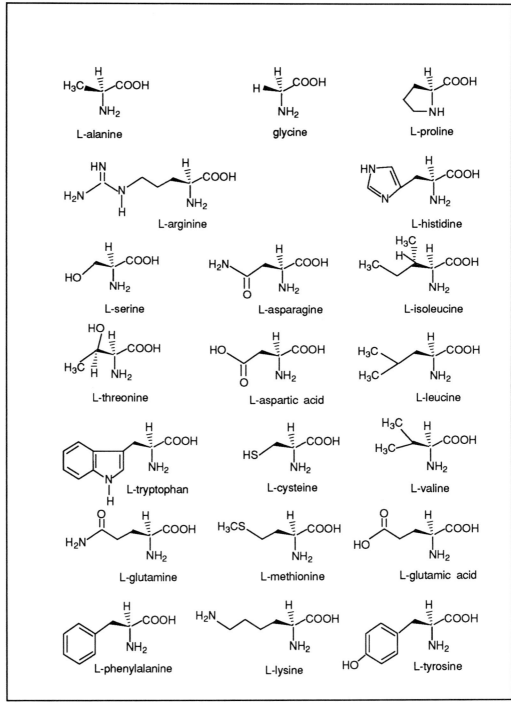

Figure 8.1 The α-amino acids found in proteins. Note that common abbreviations for the amino acids are given in the back of this text.

proteinogenic amino acids

Of interest to us here are the proteinogenic amino acids (those which occur naturally in proteins). An important fact to remember is that in proteins, amino acids usually occur as the L-isomer, whereas some micro-organisms may produce them as the D-isomer. (Figure 8.2).

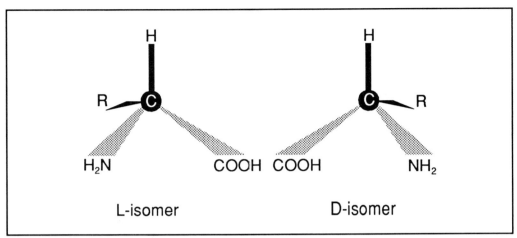

L-isomer   D-isomer

Figure 8.2 Absolute configurations of the L- and D-isomers of an amino acid.

The applications of proteinogenic amino acids in the food industry, and the production volumes of the most important ones are given in Table 8.1.

L-Amino Acid	Application	Volume (tonnes per year)
Glu (as MSG)	Flavour enhancer	270 000
Ala, Gly	Flavours	-
Asp	Manufacture of Aspartame	8 000
Phe	Manufacture of Aspartame	8 000
Lys	Dietary supplement	90 000
Cys	Reducing agent (antioxidant)	-
Thr	Dietary supplement	500
Trp	Dietary supplement	100

Table 8.1 Amino acids in the food industry (MSG = Monosodium glutamate).

∏ One amino acid is used much more than any other as a dietary supplement. Which one is it?

Lysine is used as a dietary supplement in greater quantities than other amino acids. Lysine is an essential amino acid (it cannot be synthesised by humans) and therefore must be taken in with the diet. Lysine can be added to certain foods which lack it.

There are three basic methods for amino acid production:

* fermentation;

* precursor addition;

* enzymatic synthesis.

The last two of the three methods above are also referred to as bioconversion.

We will now examine each of these methods in detail.

∏ Amino acids can be easily produced by digestion of plant or animal proteins (by acids or enzymes). What do you think the problems of producing an amino acid in this way would be?

Animal or plant proteins consist of a mixture of amino acids. The amino acid required is likely to be present in protein hydrolysates as a minor component (perhaps 5% or so of the total amino acid content). Separating this mixture to produce the required amino acid in a relatively pure form and on a large scale is a complex and expensive procedure.

8.3 Amino acid production by fermentation

Some micro-organisms, such as *E.coli*, can synthesise all the proteinogenic amino acids (and so can grow in a medium containing ammonia as the sole source of nitrogen). Other micro-organisms may require one or more amino acids as growth factors in the medium, as they are unable to synthesise them.

∏ Would you expect micro-organisms to excrete amino acids from the cell as part of their normal metabolism?

amino acid pool

Once synthesised, or taken up by the cell from the medium, amino acids are held in the cytoplasm until they are required for protein synthesis. This stock of amino acids is called the amino acid pool. It is to the advantage of the cell to retain these amino acids. It would not be to the advantage of the cell to excrete them. In amino acid fermentation, various means are therefore used to stimulate cells to excrete amino acids.

8.3.1 Use of wild-type organisms

feedback
regulation
overproduction

Normally, the synthesis of amino acids is subject to feedback regulation. The cells only produce as much as they need. However, in some instances, where the amino acid occurs in both the biosynthetic and energy-production pathways, overproduction of the amino acid can take place. We will examine this in more detail later. This is the case with L-glutamic acid in bacteria such as *Corynebacterium glutamicum*, *Brevibacterium spp.* *Microbacterium spp.* and *Arthrobacterium spp.* These organisms do not release significant quantities of amino acids into the medium, so various methods are used to increase the permeability of the cell membrane.

In these organisms the vitamin biotin and C_{16-18} saturated fatty acids are involved in the synthesis of membrane phospholipids (Figure 8.3). If phospholipid synthesis is limited, the membranes of the cells become 'leaky', and glutamic acid is released into the medium. This is achieved in two ways as follows:

- Biotin limitation. If biotin is limiting then phospholipid synthesis is limited. Biotin limitation is achieved by using, in batch culture, a medium which contains low (growth-limiting) biotin concentrations, or by using in fed-batch culture medium which lacks biotin completely and feeding biotin (or a medium containing it) at sub-optimal rates.

- Addition of C_{16-18} fatty acids. Acetyl-CoA carboxylase is repressed by high concentrations C_{16-18} fatty acids. Their addition to the medium thus has the same effect as biotin limitation.

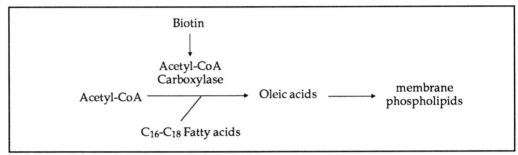

Figure 8.3 The formation of membrane phospholipids.

A third method of producing leaky cells is to add the antibiotic penicillin to the medium. At certain concentrations penicillin will cause lysis of the cells, whereas at lower concentrations the cells remain intact, but the structure of the walls and membrane become altered, and the cells become leaky.

8.3.2 Use of regulatory mutants

feedback
inhibition and
repression

The general absence of overproduction of amino acids by wild-type strains in culture media is the result of regulatory mechanisms in the biosynthetic pathway. These regulatory mechanisms are feedback inhibition and repression.

Feedback inhibition

allosteric
inhibition

In the metabolic pathway leading to the synthesis of an amino acid, several steps are involved. Each step is the result of an enzymatic activity. The key enzymatic activity (usually the first enzyme unique to the pathway) is inhibited by the end product. If the concentration of the amino acid is high, the enzymatic activity is decreased by interaction of the amino acid with the regulatory site of the enzyme (allosteric inhibition). This is called feedback inhibition.

Repression

In prokaryotic micro-organisms, enzyme production is regulated at the DNA level. The genetic information for a particular enzyme is located on the genetic material in the cell, the DNA. In general this consists of several integrated regions working together. An example is given in Figure 8.4.

These regulatory mechanisms prevent overproduction of the amino acid, and are therefore undesirable for production purposes. Mutants which lack the control mechanisms are likely to overproduce the amino acids.

∏ See if you can explain the mechanisms by which mutants would overcome feedback inhibition and repression.

Organisation of a section of DNA containing integrated regions controlling histidinol dehydrogenase sythesis.

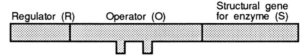

In the absence of histidine the repressor molecule cannot bind to (and 'switch off') the operator region. The operator remains 'switched on' and synthesis of the enzyme proceeds.

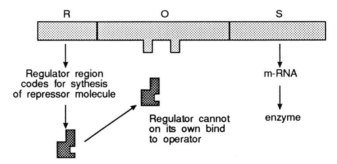

In the presence of histidine (the co-processor), the histidine binds to the regulator molecule to form a complex. This complex can bind to the operator and when so bound the operator is 'switched off'. This prevents transcription of the gene for the enzyme.

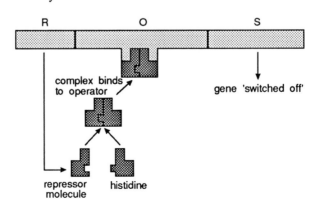

Figure 8.4 Repression of histidinol dehydrogenase.

A mutation in the gene coding for the enzyme (ie the structural gene) could lead to an enzyme with altered structure. If this altered enzyme suffers from decreased interaction with the inhibitor, then feedback inhibition can be reduced or absent altogether. A mutation in the regulator (R) gene can lead to its product (the repressor) having an altered structure. If this is unable to bind to the co-repressor, a functional repressor-co-repressor complex cannot be formed and the gene for the enzyme will remain 'switched on'.

∏ Do you think that genetic engineering could be usefully applied to the production of regulatory mutants?

randomly generated mutants/role of site-directed mutagenesis

Until recently, mutants have been generated randomly, by exposing microbial populations to mutagens. Mutants are selected by growing treated organisms up on medium containing a toxic analogue of the amino acid (usually a fluorinated derivative). Any organism which grows in these conditions is likely to have an altered regulatory mechanism. This needs an explanation. If the analogue acts as a co-repressor of a gene involved in amino acid biosynthesis, the synthesis of the normal amino acid would be switched off in the presence of the analogue. The organism would not be able to grow in these circumstances since it would be deprived of the amino acid and would use the analogue, which is toxic. If its regulator gene (and therefore repressor) had been changed so that it no longer bound the co-repressor or did not bind with the operator, then the organism would be no longer sensitive to the amino acid analogue. Such a strain would not be subject to feedback repression. There is, therefore, no need to resort to recombinant DNA technology to produce desired mutant strains. Nevertheless, once a mutant gene has been produced, genetic engineering could be used to transfer it to other organisms. Furthermore, site-directed mutagenesis is a technique which could be applied to the production of the required mutations in a controlled manner since it enables sequences of DNA to be altered in a predetermined manner. DNA technology can also be used to generate strains that over-produce amino acids by using strong promoters to control the relevant genes.

8.3.3 Use of auxotrophic mutants

auxotrophic mutants

Auxotrophic mutants (ones which require growth factors which the parental organism does not) are useful in the production of amino acids. As an example, consider the pathway for the production of L-lysine by *Corynebacterium glutamicum* (Figure 8.5).

The common intermediate, aspartic acid, is converted via aspartate semi-aldehyde to either lysine or homoserine. Homoserine is itself converted to methionine and threonine. The enzyme aspartate kinase suffers feedback inhibition from both L-lysine and L-threonine. Both are required for complete inhibition of this enzyme.

∏ If a mutant organism had a defective (inactive) homoserine dehydrogenase enzyme, what effect do you think this would have on the levels of lysine produced? (Figure 8.5 will help you answer this).

Such a mutant would be unable to convert aspartate semi-aldehyde to homoserine. Thus all the aspartate semi-aldehyde would be shunted into lysine formation and so the levels of lysine produced would be raised.

An organism with defective homoserine dehydrogenase cannot form methionine or threonine and so these growth factors must be added to the growth medium. The

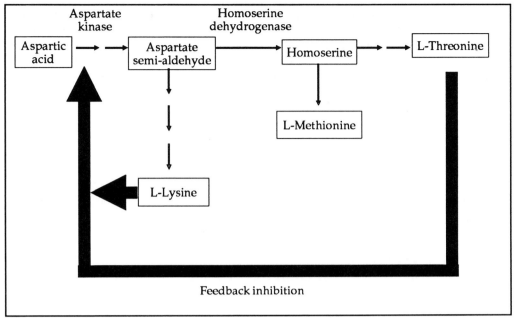

Figure 8.5 Regulation of L-lysine biosynthesis in *Corynebacterium glutamicum*.

organism is a threonine/methionine auxotroph and is written as Δ Thr, Δ Met (or Thr⁻, Met⁻). In this text we will use the latter nomenclature. The organism requires either threonine and methionine or homoserine for growth. These must however, be added at low concentrations, otherwise threonine will inhibit aspartate kinase and so reduce the levels of lysine produced. In practice, the growth factors are added in a fed-batch culture system at a rate which is just less than the rate at which the culture would otherwise use them (in other words, they are added at sub-optimal rates, so that they are used up immediately and so do not accumulate in the medium).

SAQ 8.1

Answer true or false to each of the following:

1) Use is made of mutants in amino acid production because wild-type organisms do not excrete amino acids in measurable quantities.

2) Auxotrophic mutants overproduce amino acids because they do not suffer from feedback inhibition.

3) Regulatory mutants may overproduce amino acids because they are freed from the regulatory effect of the amino acid products.

4) Auxotrophic mutants may overproduce amino acids because intermediates are diverted towards synthesis of the required amino acid instead of being used in synthesis of other amino acids.

SAQ 8.2

The following diagram represents the metabolic pathway leading from an intermediate A to three amino acids - E, H and J.

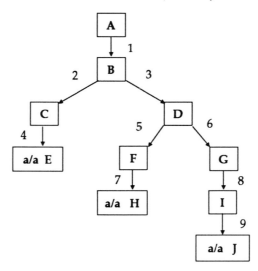

The required product is the amino acid H. Note that:

E exerts feedback inhibition on enzyme 1;

J exerts feedback inhibition on enzyme 6.

Use these data to answer the following questions.

1) Which of the following would be most likely to overproduce H?

   a) Mutant 1 E⁻   (defective enzyme 2).

   b) Mutant 2 J⁻   (defective enzyme 8).

   c) Mutant 3 E⁻ H⁻   (defective enzymes 4 and 5).

   d) Mutant 4 J⁻ H⁻   (defective enzymes 7 and 9).

   e) Mutant 5 E⁻ J⁻   (defective enzymes 2 and 8).

   f) Mutant 6 E⁻ H⁻ J⁻   (defective enzymes 2 and 3).

2) For the organism chosen in 1), as being most likely to overproduce H, which of the following amino acids would be required in the production medium: E; H; J?

3) For the amino acid(s) chosen in 2), would production of H be lowered if it (they) were present at relatively high concentrations?

8.3.4 The fermentation process

stirred aerated fermenters

Fermentation processes for amino acids are performed in conventional stirred, aerated fermenters (Figure 8.6). Fermentations are aseptic and the fermenter is equipped with temperature, pH and foam control. Vessels may be up to a few hundred m³, depending on the quality of product required.

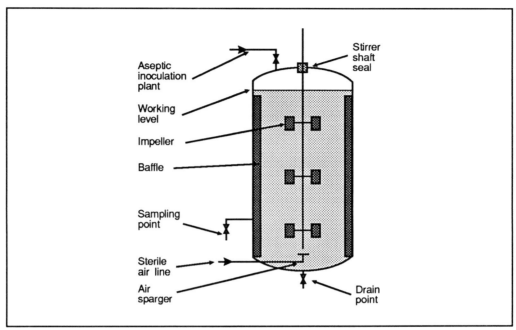

Figure 8.6 Stirred fermenter. The one shown is fitted with baffles to increase mixing.

∏ It may be obvious, but can you explain why amino acid fermentations have to be operated under aseptic conditions? (Before reading on, see if you can write down 3 or 4 reasons).

Contamination can lead to lower yields of amino acids (as the contaminants compete with the production organism for the substrate). Contaminants may produce metabolites (such as organic acids) which inhibit growth of the production organism and/or product formation. The contaminant may have a requirement for the amino acid being produced and so will use some of it up as it grows. The contaminant may produce enzymes such as deaminases or decarboxylases which convert the amino acid into other compounds. For these reasons, aseptic operation is essential.

batch cultures   Batch cultures have been traditionally used for amino acid production. Amino acid production occurs during the growth phase and the stationary phase. As with all industrial fermentation processes, the lag phase (which can occur after inoculum addition) must be kept to a minimum as this wastes time and hence money.

∏ How do you think lag phases are kept to a minimum in industrial fermentations?

The lag phase (the time taken for an inoculum to start to grow) occurs because the organisms in the inoculum are readjusting to a new growth environment. If the inoculum is prepared in medium having the same compositions as the production medium and is transferred while still in the phase of active growth, then lag phases will be kept to a minimum. An inoculum level of 5-10% (v/v) is normally used. It is worth remembering the general rules, the larger the inoculum, the shorter the lag phase and the younger the inoculum, the shorter the lag phase.

large, young inocula

fed-batch cultures

Fed-batch cultures offer advantages over batch cultures. One or more nutrients are fed to the culture, either continuously or intermittently. If the nutrients are fed at sub-optimal rates (at a rate which is less than the rate at which the culture could use the nutrients) then the nutrients will not accumulate in the medium. This can overcome any inhibitory effects of medium components (such as repression), or can be used to stimulate excretion of amino acids from the cells (see Section 8.3.1). It can also reduce the growth rate and thus the oxygen demand (the quantity of oxygen required by a culture). In other words, it can ensure that oxygen does not become depleted during the phase of exponential growth when the oxygen demand is high.

sub-optimal rates

continuous and semi-continuous cultures

Continuous or semi-continuous cultures have the advantage of higher output for a given reactor size. Installation and maintenance costs are reduced because of the smaller volume required. In continuous cultures, the environment remains constant throughout (as opposed to batch or fed-batch cultures where the environment changes throughout the growth cycle).

∏ List 3 disadvantages that may be encountered in producing amino acids in continuous culture. (Think back to the problems facing the long-term maintenance of starter cultures used in cheese manufacture discussed in Chapter 4).

Continuous cultures can be prone to contamination, which has the disadvantages we outlined for batch cultures. A more serious disadvantage is the increased likelihood of loss of desirable characteristics either by mutation (back mutation) or loss of plasmids. In addition, phage infections can arise.

In practice, continuous culture is applied to glutamic acid production by *Microbacterium ammoniaphilum* in a multi-stage system. In this system the contents of the first fermenter is fed (continuously) into another vessel (or vessels). The flow rate in the second vessel is lower than the first, and so the residence time (the time the medium spends in the vessel) is longer. In this way all the substrate is completely used up.

8.3.5 The problems in amino acid fermentation

The major problems in amino acid fermentation are:

* contamination during fermentation;

* variability in quality of the substrates;

* infection by bacteriophages;

* back mutation or loss of genetic material.

reducing problems

The first two items can be prevented respectively by proper plant design and operation and pooling of large stocks of standardised ingredients. Phage infection is prevented by suitable sterilisation procedures (including the air introduced into the fermenter). It can be reduced by operating the process under unfavourable conditions (for example, at a temperature at which the phage cannot replicate). Alternatively, an agent can be included in the medium which prevents phage infection, such as a chelating agent which sequesters calcium ions and so prevents phage attachment to host cells.

genetic stability   Genetic stability is ensured by preparing stock cultures, from which fresh inocula are regularly made (often for each run). If the organism is a recombinant carrying plasmids, their loss can be prevented by including the antibiotic in the medium for which the plasmid carries the resistance gene. An improvement in gene stability is usually achieved if the gene is incorporated into the host cell chromosome.

Π   Explain how, if the plasmid carries a gene for antibiotic resistance, putting the antibiotic in the medium ensures that the plasmid is not lost by the culture.

In a normal medium (without antibiotic), plasmid loss will not hinder the organism's growth and in some circumstances may improve growth. If plasmid loss leads to higher growth rates, this organism will eventually predominate in a culture. However, if an antibiotic, the resistance gene for which is coded for on the plasmids, is put into the medium, any organism which loses a plasmid and so becomes sensitive to the antibiotic will be unable to grow.

Phage infection and genetic instability will be a greater problem in semi-continuous or continuous cultures, because of the long culture periods involved.

| SAQ 8.3 | Which of the following support amino acid fermentation by fed-batch culture, rather than batch culture? |

1) Fed-batch culture allows the growth rate and oxygen demand of a culture to be controlled.

2) Fed-batch cultures have low probability of phage infection, due to the short growth period.

3) Fed-batch cultures have low probability of genetic reversion, due to the short growth period.

4) Fed-batch cultures allow amino acids required by auxotrophic mutants to be added at sub-optimal levels so as to avoid feedback inhibition.

5) Fed-batch cultures allow manipulation of the medium composition so as to stimulate excretion of amino acid from the cell.

8.3.6 Down-stream processing in amino acid fermentations

As with all fermentation processes, the downstream processing should:

* be as simple as possible;

* have a high recovery yield (recovery yield is the amount of product recovered divided by the amount of product present in the fermentation broth);

* avoid any steps causing loss of product;

- be economic;

- be easy to scale up to large scale;

Amino acid recovery does have some advantages over other processes such as enzyme production. The overproduction of the required amino acid means that the concentration of unwanted (contaminating) amino acids is relatively low. If only the L-form of the amino acid is synthesised, resolution (separation) of the D-isomer is unnecessary and protein impurities are relatively easy to separate and remove.

The first stage is cell removal which is achieved by filtration (usually by rotary vacuum filters, plate-frame filters or pressure leaf filters). Alternatively, centrifugation, using various types of continuous centrifuges can be used. Sometimes flocculating agents such as aluminium hydroxide, or filtration aids, such as diatomaceous earth, need to be added to improve the performance of centrifuges and filters respectively. Once cells have been removed, the proteins in the medium are removed, by ultrafiltration or precipitation. These steps are necessary as cells and/or proteins would interfere with amino acid recovery procedures. Several methods of recovery of amino acids from the cell-free deproteinised medium are practiced.

Crystallisation

Crystallisation of amino acids is possible as the solubility of the (amphoteric) product is affected by pH and is usually minimum at the isoelectric point. The isoelectric point is the pH at which the amino acid carries no net electrostatic charge. Adjustment of pH, lowering the temperature and addition of ammonium or calcium salts will cause the amino acid to crystallise and precipitate out as its salt. The precipitate can be recovered by centrifugation and treated with acid or alkali to liberate the free amino acid from the salt.

Ion exchange

Ion exchange resins have been widely used in the recovery of amino acids from fermentation broths. They are available in the form of cationic resins (which bind positively charged molecules) and anionic resins (which bind negatively charged molecules).

Π Explain, in outline, how you would adjust the medium pH in order to use a resin to recover a particular amino acid. Do this in the form of a flow diagram and then see if it fits with our description below.

When the pH is lower than the isoelectric point, the amino acid will have positive charge (as the amino group ionises) and will bind to a cationic resin. If the pH is higher than the isoelectric point, the amino acid has negative charge (as the carboxyl group ionises) and will bind to an anionic resin. By passing the medium through a bed (column) of appropriate resin at appropriate pH, the amino acid can be adsorbed onto the resin. By reversing the pH (from below the isoelectric point to above it, or vice versa) the amino acid can be eluted. Amino acids can also be eluted from such resins by using salt solutions.

Resins of this type are not widely applied to amino acid recovery from fermentation broths as they are relatively expensive and difficult to operate on a large scale.

Once the amino acid is recovered, coloured impurities are usually removed. This is carried out by washing the crystalline product in water (using a pH at which the crystals do not dissolve) or by passing the solution through a bed of activated charcoal.

SAQ 8.4

Answer true or false to each of the following:

1) In amino acid recovery from fermentation broths, the broths are concentrated by vacuum drying to cause crystallization of the product.

2) The solubility of an amino acid in a fermentation broth is greatest when the pH of the broth coincides with the isoelectric point of the amino acid.

3) The first stage in downstream processing for amino acid recovery is the removal of biomass, as the biomass would interfere with subsequent recovery processes.

4) A cationic resin can be used to recover an amino acid when the pH of the medium is lower than the isoelectric point of the amino acid.

8.4 Amino acid production by bioconversions

In the previous section, we examined amino acid production by fermentation. While fermentation has been established for several amino acids, it has not been used for others (including some of interest in the food industry, particularly alanine, cysteine and aspartic acid). For these amino acids bioconversions have been or are being

intermediates   developed. This involves using microbial cells or their enzymes to convert either an intermediate in the synthetic pathway leading to the amino acid, or a related compound not normally an intermediate in the pathway (called a precursor) to the amino acid. Let

precursors   us examine these processes in more detail.

8.4.1 Precursor addition

Precursors, which may be chemically synthesised, are normally similar in structure to the amino acid, but are not intermediates in normal metabolic pathways. In this way the precursor does not regulate amino acid production (by feedback inhibition or repression). The following example illustrates this.

L-threonine   The enzyme L-threonine hydratase catalyses the conversion of L-threonine to
hydratase   α-ketobutyrate, an intermediate in the formation of L-isoleucine (Figure 8.7). The enzyme is subject to feedback inhibition by isoleucine, so addition of L-threonine or α-ketobutyrate does not cause overproduction of isoleucine. However, if D-threonine is
D-threonine   added as a precursor, the enzyme D-threonine hydratase converts D-threonine to
hydratase   α-ketobutyrate. The enzyme can be induced by D-threonine and it does not suffer from feedback inhibition by L-isoleucine. Thus, overproduction of isoleucine occurs.

L-isoleucine can also be produced by using α-bromobutyrate, α-hydroxybutyrate or α-aminobutyrate as precursor (in place of α-ketobutyrate as an intermediate).

The precursor may be added directly to a culture, or after the cells have been concentrated by centrifugation and suspended in a smaller volume of fresh medium (ie a concentrated cell suspension).

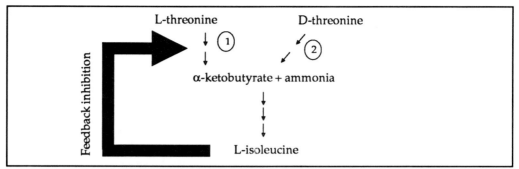

Figure 8.7 Production of L-isoleucine by precursor addition. 1: L-threonine hydratase, 2: D-threonine hydratase.

If a batch system is used, the precursor is added to the cell culture or suspension and once conversion has taken place the cells must be separated to allow product recovery and possible re-use of cells. If continuous cultures are used the cells would tend to become diluted.

Π   Describe at least one continuous system in which the cell concentration could be maintained at artificially high levels.

continuous culture

biomass feedback

downstream

If your answer is continuous culture with biomass feedback, this is correct, although such a system would be technically difficult to operate. Such systems could be operated in one of two basic ways. In the first way, the outflow from the vessel is centrifuged or filtered to separate cells from culture medium. The cells are returned to the reaction vessel (external feedback). The other way would be to fit a filter on the outflow duct so that only media could be removed from the vessel (internal feedback). Using immobilised cells in a continuous reactor system is another way of operating continuous systems. It also has the advantage of producing a product stream with a low concentration of biomass, which makes downstream processing easier.

8.4.2 Enzymatic synthesis

The use of enzymes for production of amino acids has been stimulated by development of immobilisation techniques, which allows the use of convenient and efficient continuous reaction processes. We will not consider enzyme immobilisation methods in detail here. You will find them described in the BIOTOL text 'Enzyme Technology'. Briefly the methods are:

* covalent attachment to solid supports;

* adsorption onto solid supports;

* entrapment in polymeric gels;

* cross-linking of enzyme molecules;

* encapsulation.

Each method has relative advantages and disadvantages, which must be considered when a process is being developed. The major reaction system is the flow-through packed-bed column reactor (see Figure 6.17).

As our knowledge of mechanisms of enzyme specificity and thermostability increases, so does the possibility of improving enzyme performance, using modern techniques of gene manipulation and protein engineering. It is also recognised that enzymes may show desirable activities in non-aqueous media. In general use of enzymic reactions in reactors is limited to those enzymes (such as hydrolases or isomerases) which do not require expensive cofactors such as ATP, NAD$^+$ or NADP$^+$. Let us examine how enzymatic synthesis operates.

hydrolases
isomerases

8.4.3 Production of L-amino acids by enzymatic resolution of racemic mixtures

It is often possible to produce amino acids from racemic mixtures (mixtures of both D- and L-amino acids) or substituted amino acids, which have been produced by fermentation and/or by chemical synthesis.

The Japanese firm Tanabe Inc Ltd has been operating, since 1969, the resolution of D, L-amino acids using aminoacylase. The principle is based on the asymmetrical hydrolysis of N-acyl-D,L-amino acid by aminoacylase which gives the L-amino acid and the unhydrolysed acyl-D-amino acid (Figure 8.8).

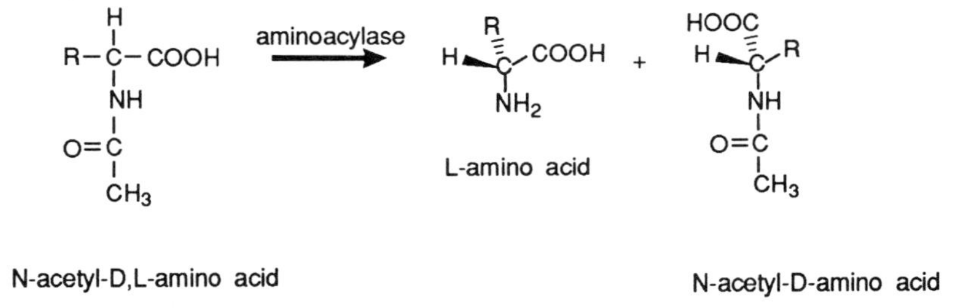

Figure 8.8 Process for the enzymatic production of L-amino acids from N-acetyl-D,L-amino acids.

methods of
enzyme
immobilisation
evaluated

To develop a continuous process, the immobilisation of aminoacylase of *Aspergillus oryzae* by a variety of methods was studied. These included ionic binding to DEAE-Sephadex, covalent binding to iodo-acetyl cellulose and entrapment in polyacrylamide gel. Ionic binding to DEAE-Sephadex was chosen because the method of preparation was easy, activity was high and stable, and the support could be re-used.

The flow diagram of the enzyme reactor for continuous production of L-amino acids is given in Figure 8.9. The N-acetyl-amino acid is continuously run into the enzyme column through a filter and a heat exchanger. The effluent is concentrated and the L-amino acid is crystallised. The N-acyl-D-amino acid contained in the mother liquor is racemised by heating in a racemisation tank and re-used.

∏   What three physical parameters of the enzyme column are controlled in the process illustrated in Figure 8.9.

The flow rate, the temperature and the pH.

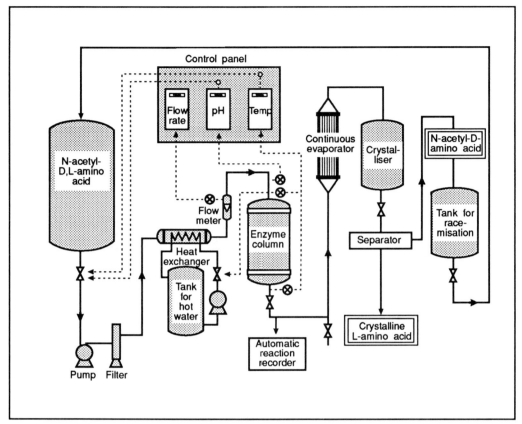

Figure 8.9 Flow diagram for the continuous production of L-amino acids by immobilized aminoacylase.

The advantages of the continuous process over a batch process are:

- saving of enzyme;

- saving of labour, due to automation;

- increase in reaction yield, due to the easy isolation of L-amino acid from the reaction mixture.

As a result of these advantages, the overall production costs are 40% lower than that of the batch process. A comparison between the batch process costs and the continuous production costs is given in Figure 8.10.

Π   Does immobilisation and a continuous process produce the largest reduction in 1) the cost of labour, 2) enzyme or 3) substrate, compared to batch processes?

In the continuous process the cost of enzyme is reduced to about 1/10 of the batch cost, while the cost of labour is reduced to about 1/3 of the batch cost. The cost of the substrate is similar in both processes. The greatest reduction in cost is therefore in the cost of the enzyme.

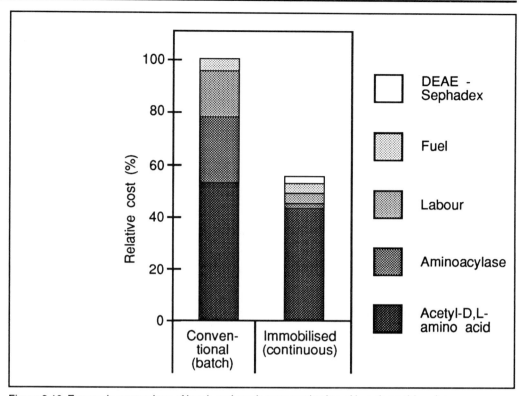

Figure 8.10 Economic comparison of batch and continuous production of L-amino acids using aminoacylases.

The plant is mainly used for the amino acids L-methionine, L-valine and L-phenylalanine. The German firm, Degussa A.G. uses immobilised aminoacylase to produce a variety of L-amino acids, for example L-methionine. The principles of the process are the same as the Tanabe-process. Degussa uses a new type of reactor, an enzyme membrane reactor, on a pilot plant scale to produce L-methionine, L-phenylalanine and L-valine in an amount of 200 tonnes per year (Figure 8.11). The substrate is pumped through a sterilising filter into the reactor. The reactor contains native (soluble, unbound) enzyme and the reaction mixture is recycled around the system. The reactor contains an ultrafiltration membrane, which allows the product molecules to pass through, while retaining the enzyme. Product formation is continuously monitored by polarimetry and additional enzyme added as required.

> ∏ Make a list of the advantages of the enzyme membrane reactor, compared to a continuous immobilised enzyme reactor (such as that described in Figure 8.9). (See if you can think of at least three before you read on).

The fact that the enzyme is not immobilised saves costs (the cost of labour and the support material). This also overcomes some potential problems associated with immobilisation, such as loss of activity or stability on immobilisation and diffusion limitation (whereby the support restricts access of the substrate to the enzyme). The system also gives a higher output per unit volume and additional enzyme can be easily added to the system during operation.

enzyme membrane reactor

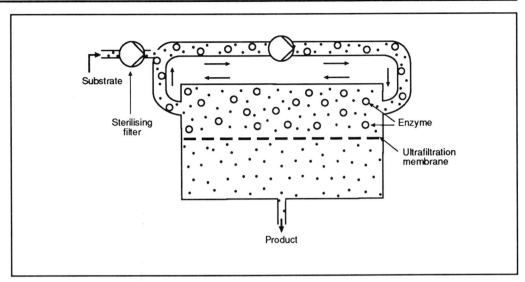

Figure 8.11 The enzyme membrane reactor.

The disadvantages of the enzyme membrane reactor are that investment (installation) costs are high, and enzymes are prone to deactivation within the system (partly as a result of sheer forces due to recycling around the system).

α-amino-ε-caprolactam hydrolase

α-amino-ε-caprolactam racemase

Another process to resolve racemic mixtures is used in the production of L-lysine by Toray Industries. D, L α-amino-ε-caprolactam is obtained by chemical synthesis from cyclohexene. This racemic (D,L) mixture is hydrolysed by L-specific α-amino-ε-caprolactam hydrolase from the yeast *Candida lumicole* (Figure 8.12). This produces L-lysine from the L-isomer. The D-isomer is racemised by α-amino-ε-caprolactam racemase from the bacterium *Alcaligenes faecalis* into a racemic mixture which is recycled.

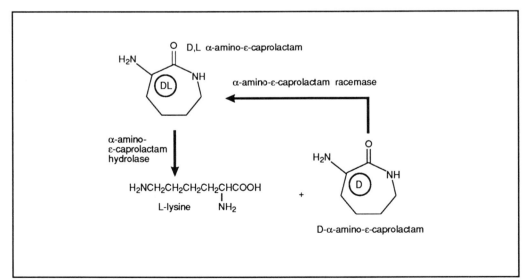

Figure 8.12 The production of optically pure L-lysine from α-amino-ε-caprolactams.

At DSM a very efficient and universal method has been developed for the production of optically pure L- and D-amino acids. This process is at the moment operated on pilot plant scale and uses amidase activity. The principle is based on the enantioselective hydrolysis of D,L-amino acid reaction conditions starting from simple raw materials (Figure 8.13). Thus reaction of an aldehyde with hydrogen cyanide in ammonia (Strecker reaction) gives rise to the formation of the aminonitrile. The aminonitrile is converted in high yield to the D,L-amino acid amide under alkaline conditions in the presence of a catalytic amount of acetone.

The resolution step is accomplished with permeabilised whole cells of *Pseudomonas putida* ATCC 12633 and a nearly 100% stereoselectivity in hydrolysing only the L-amino acid amide is combined with a very broad substrate specificity (see Table 8.2).

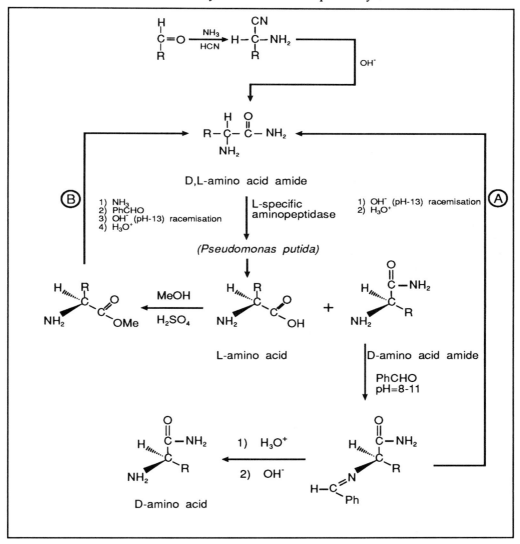

Figure 8.13 DSM-process for the production of optically pure L- and D-amino acid. Path A and B are explained in the text.

Not only the smallest optically active amino acid ie alanine, but also leucine, several (substituted aromatic amino acids, heterosubstituted amino acids (methionine, homomethionine and thienylglycine) and even an imino acid, proline are obtainable in both the L- and D-form.

No enzymic side effects are observed and substrate concentrations up to 20% by weight can be used without affecting the enzyme activity. The biocatalyst is used in soluble form in a batchwise process, thus poorly soluble amino acids can be resolved without technical difficulties. Re-use of the biocatalyst is in principle possible. A very simple and elegant alternative to the use of ion-exchange columns or extraction to separate the mixture of D-amino acid amide and the L-amino acid has been elaborated at DSM. Thus addition of one equivalent of benzaldehyde (with respect to the D-amino acid amide) to the enzymic hydrolysate results suprisingly in the formation of a Schiff base with the D-amino acid amide which is insoluble in water and therefore can be easily separated. Acid hydrolysis (H_2SO_4), HNO_3, etc) results in the formation of the D-amino acid (without racemisation).

Table 8.2 Substrate specificity of the amino peptidase from *Pseudomonas putida* ATCC 12633 (selected examples of substrates used by the enzyme).

Alternatively the D-amino acid amide can be hydrolysed by a cell preparation of *Rhodococcus erythropolis*. This biocatalyst lacks stereoselectivity. This option is very useful for amino acids which are highly soluble in the neutralised reaction mixture obtained after acid hydrolysis of the amide.

Process economics dictate the recycling of the unwanted isomer. Path A in Figure 8.13 illustrates that racemisation of the D,N-benzylidene amino acid amide is facile and can be carried out under very mild reaction conditions. After removal of the benzaldehyde, the D,L-amino acid amide can be recycled. This option shows that 100% conversion to the L-amino acid is theoretically possible. A suitable method for racemisation and recycling of the L-amino acid (path B, Figure 8.13) comprises the conversion of the L-amino acid into the ester in the presence of concentrated acid, followed by addition of ammonia, resulting in the formation of the amide. Addition of benzaldehyde and racemisation by OH⁻ (pH = 13) gives the D,L-amino acid amide. In this way 100% conversion to the D-amino acid is possible. The presence of an α-hydrogen in the substrate is an essential structural feature for the enzymic activity of the aminopeptidase of *Pseudomonas putida*.

Enantiomerically pure α-disubstituted amino acids, on the other hand, are interesting, mainly because of their activity as enzyme-inhibitors. To provide access to this class of amino acids a new biocatalyst from *Mycobacterium neoaurum* has been developed at DSM capable of stereoselectivity hydrolising a range of α-C-disubstitued amino acid amides (Figure 8.14).

The resolution method is based on a reaction scheme similar to the one developed for the production of α-L-amino acids (Figure 8.13).

It is impossible to racemise α-disubstituted amino acids. Recycling of the unwanted isomer can only be achieved, therefore, by converting the amino acid into the starting material (ketone). Compared to some other enzymic resolution methods (especially those based on acylases and esterases, the apporaches described so far show a distinct advantage in that the substrate for the enzymatic hydrolysis is a precursor of the amino acid. Therefore the number of chemical steps can be kept to a minimum. Furthermore the use of relatively cheap whole cell biocatalysts contributes to the economical feasibility of these procedures.

L- as well as D-amino acids can be prepared with a very high optical purity.

whole cells
Instead of using purified immobilised enzymes to carry out these reactions, the process uses immobilised whole cells (think of the cells as being sacs of enzymes). In this way production costs are reduced and the process is essentially the same as precursor addition.

Π   Divide a piece of paper into two columns. In one column write down the disadvantages the use of whole cells might have, compared to using cell-free enzymes. In the other column, write down the advantages of using whole cells.

The main disadvantage we can identify is that whole cells contain a wide range of enzymes, in addition to the required one. This might adversely affect the product. For instance, if a cell contains the D-amino acid racemase, some of the L-amino acid product may be converted to unwanted D-amino acid. Other products (such as organic acids) may alter the pH of the reaction mixture and affect enzyme activity. Another disadvantage might be that the substrate might have only limited access to the relevant enzyme with the cell (ie the reaction might be slower). The enzyme may be unstable in whole cells. The advantages are that the use of whole cells saves on the costs of immobilisation, the enzyme may be stable in whole cells and if the enzyme requires cofactors the cell can supply (and regenerate) them.

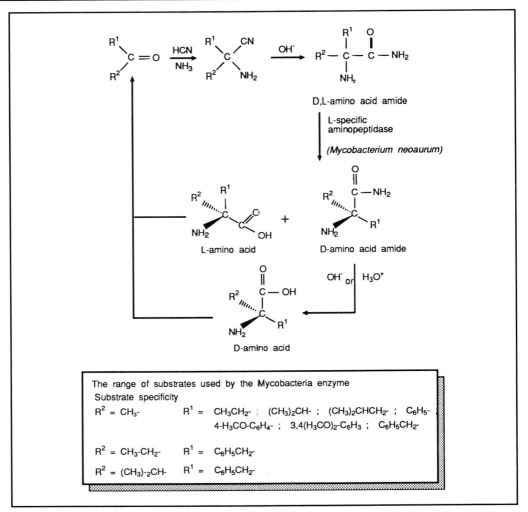

Figure 8.14 Enzymatic method for the optical resolution of racemic α-disubstituted amino acids.

ATC racemase
L-ATC
hydrolase
S-carbamoyl-L-
cysteine
hydrolase

A racemic mixture of chemically synthesised D,L-2-amino-Δ²-thiazolin-4-carboxylate (D,L-ATC) can be resolved and converted to L-cysteine by cells of *Pseudomonas putida* AJ3854 (Figure 8.15). This organism contains the enzyme ATC racemase which converts D-ATC to L-ATC. The enzyme L-ATC hydrolase converts L-ATC to S-carbamoyl-L-cysteine, which is in turn converted to L-cysteine by the enzyme S-carbamoyl-L-cysteine hydrolase.

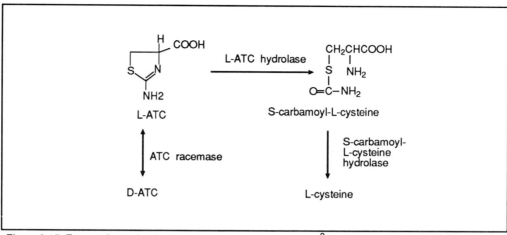

Figure 8.15 Enzymatic synthesis of L-cysteine from D,L-2-amino-Δ^2-thiazolinon-4-carboxylate (D,L-ATC).

8.4.4 Production of L-amino acids using ammonia lyases

phenylalanine
ammonia lyase

L-Phenylalanine can be enzymatically synthesised from trans-cinnamic acid (Figure 8.16). A process has been operated by Genex, using the enzyme phenylalanine ammonia lyase from the yeast *Rhodotorula rubra*.

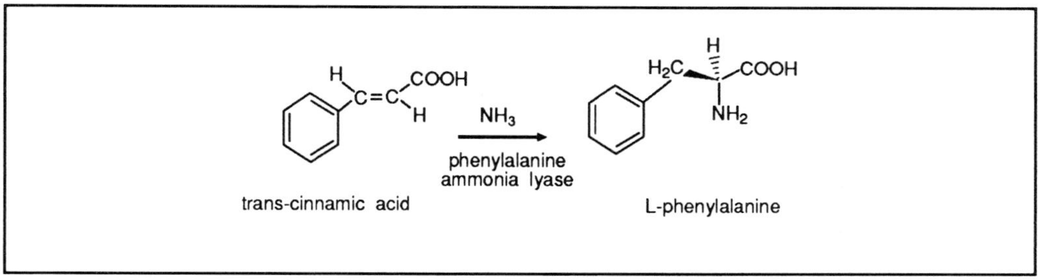

Figure 8.16 Enzymatic conversion of trans-cinnamic acid to L-phenylalanine.

Similarly, fumaric acid can be converted to L-aspartic acid by aspartase (Figure 8.17). Continuous processes have been developed using immobilised aspartase from *E. coli*. Conversion levels of 99% have been achieved and the enzyme system seems to be stable for at least 3 months. L-aspartic acid can itself be converted to L-alanine by L-aspartate-β-decarboxylase. The enzyme has been obtained from the bacteria *Xanthomonas oryzae*, *Pseudomonas dacunhae* and *Achromobacter pestifer*.

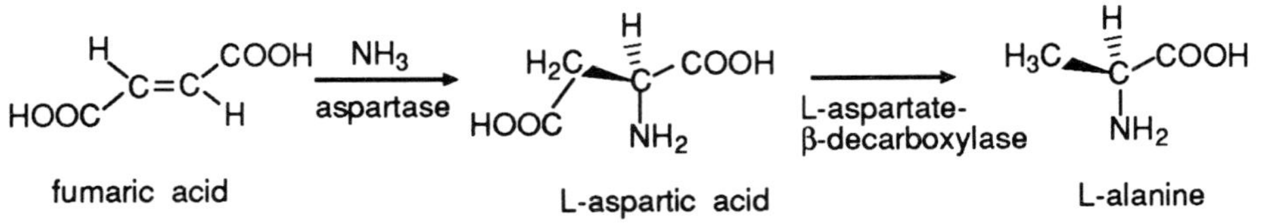

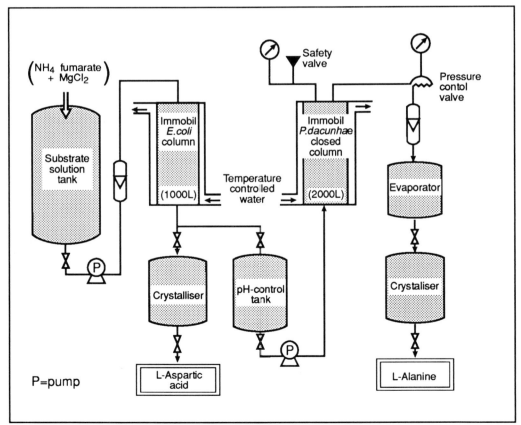

Figure 8.17 Two step enzymatic synthesis of L-alanine from fumaric acid.

aspartase

half life of
enzymes

L-aspartic acid has been produced in Japan by the Tanabe Seiyaku Co Ltd, in a batch process using whole cells of *E. coli* with high levels of aspartase activity. Subsequently this process was replaced with enzymes immobilised in beads and used in a continuous process in a packed bed reactor (Figure 8.18). Polyacrylamide was initially used as a gelling agent and it resulted in half-life times for aspartase activity of about 120 days at 37°C, (ie the aspartase activity was half of its original value after 120 days). Then, in 1978, κ-carrageenan was substituted as a gelling agent as this results in half lives of about 680 days. The column has a bed volume of about 1000 l (1m³) and is equipped with a cooling jacket (cooling is required as the reaction is exothermic). Ammonium fumarate is used instead of fumaric acid and ammonia.

Figure 8.18 Flow diagram for the continuous production of L-aspartic acid and L-alanine.

L-aspartate-β-
decarboxylase

Similarly, L-alanine was initially produced (in 1965) in batch processes using purified L-aspartate-β-decarboxylase. Later on a continuous process was developed using cells of *Pseudomonas dacunhae* immobilised in κ-carrageenan beads. A problem with this system is that the cells give off carbon dioxide.

Π What problems do you think that CO_2 production causes in such an immobilised cell system and how do you think it could be overcome? (It is probably quite easy to think of the problems that might arise - solving them is a lot more difficult - see if you can come to a conclusion before reading on).

CO_2 evolution in gel beads can lead to the rupture of the beads by gas bubble formation. Gas bubbles rising up the column cause mixing, so the system does not behave in the required manner (that is as a plug-flow system). Bubble formation can be prevented by operating the system under positive pressure, so that the CO_2 produced dissolves in the medium.

closed-column
reactor

The reactor used in the production of L-alanine is a closed-column reactor (Figure 8.19), operated at overpressures of about 4 bar (4×10^5 N m^{-2}). CO_2 remains dissolved in the reaction mixture and this improves the efficiency of the system by over 40%.

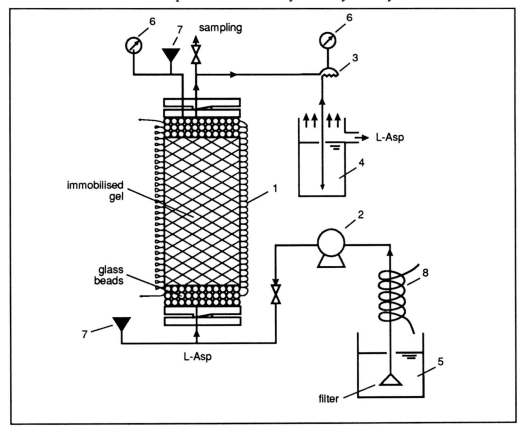

Figure 8.19 Closed-column reactor for the production of L-alanine. 1) Reactor, 2) Plunger pump, 3) Pressure control valve, 4) Reservoir, 5) Substrate tank, 6) Pressure gauge, 7) Safety valve, 8) Heat exchanger.

In order to prevent the formation of unwanted by-products such as L-malic acid and D-alanine, the cells are treated with extreme pH, in order to inactivate alanine racemase enzyme.

SAQ 8.5

Answer true or false to each of the completions to the following statements.

In the continuous production process for L-aspartic acid and L-alanine using immobilized whole cells:

1) *E. coli* has to be used in a closed-column reactor to prevent disruption of flow by CO_2 evolution;

2) the activity required in the cells of *P. dacunhae* is L-aspartate-β-decarboxylase;

3) ammonia or ammonium fumarate is added to immobilized *P. dacunhae* to act as amino donor;

4) the activity required in the cells of *E. coli* is aspartase.

SAQ 8.6

Answer true or false to each of the following statements.

1) A racemic mixture of D,L-aspartic acid can be resolved into L-aspartic acid by the appropriate amino-acylase enzyme.

2) A racemic mixture of D,L-lysine can be resolved into L-lysine by α-amino-ε-caprolactam racemase.

3) Ammonia lyase enzymes require other amino acids as amino donors.

4) L-aspartic acid can be produced from L-alanine by the action of L-aspartate decarboxylase.

SAQ 8.7

Which of the following support biotransformation using immobilized enzyme rather than free (unimmobilized) enzyme?

1) Immobilized enzymes may retain activity for several weeks.

2) Immobilized enzymes can resolve racemic mixtures.

3) Immobilized enzymes can be easily reused or used in continuous reactors.

4) Immobilized enzymes can produce a product stream free of (contaminating) enzyme.

SAQ 8.8

Choose the correct completion to the following statement.

The difference between precursor addition and other bioconversions, using whole cells, is that:

1) precursor addition uses enzymes within the cells to carry out the conversion;

2) precursor addition involves conversion of a compound which is not an intermediate in the metabolic pathway;

3) the precursor can be a chemically synthesized compound;

4) precursor addition can be carried out using immobilized cells;

5) precursor addition does not require the addition of cofactors.

8.5 L-Phenylalanine production - A case study

So far we have examined some of the methods which have been developed for the production of amino acids. In this case study we are going to examine how these methods can be applied to the production of L-phenylalanine and how these processes compare economically.

8.5.1 The uses of L-phenylalanine

Phenylalanine is used as a dietary supplement (being an essential amino acid) and in intravenous infusions. By far its largest use is in the production of the artificial sweetener aspartame. In aspartame production, the methyl ester L-phenylalanine and L-aspartic acid are chemically combined to form L-aspartyl-L-phenyalanyl-methyl ester, which is the chemical name for aspartame.

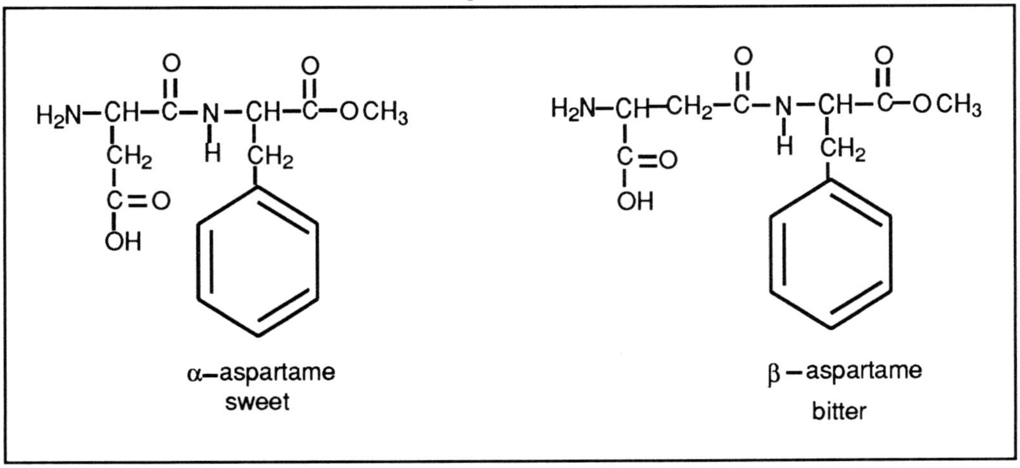

Figure 8.20 The structure of aspartame.

It is important in aspartame production that L-amino acids alone (and not D-amino acids) are incorporated into aspartame. We remind you that α-aspartame produced from L-Asp and L-Phe has a sweet taste whereas aspartame produced from D-Asp and L-Phe has a bitter taste (Figure 8.20). β-Aspartame which results from the coupling of the amino group of L-Phe with the β-carboxylic acid group of L-Asp also has a bitter taste. This is a good example of the value of being able to resolve racemic mixtures or to produce one particular isomeric form.

8.5.2 The routes to L-phenylalanine production

Before we look in detail at the technologies and economics of some of the ways of producing phenylalanine, let us review them. Below we have listed some examples.

You should appreciate that the methods listed are not the only ones available for phenylalanine (and other amino acid) production. However, they are the methods which are applied on a large scale or have been considered for development. In this case study we will compare phenylalanine production by fermentation, by precursor addition and by enzymatic synthesis from trans-cinnamic acid.

a) Fermentative production of L-phenylalanine from glucose

b) Production of L-phenylalanine by precursor addition (phenylpyruvic acid (PPA))

c) Production of L-phenylalanine from trans-cinnamic acid (TCA) and ammonia

d) Optical resolution of N-acetyl-D,L-phenylalanine

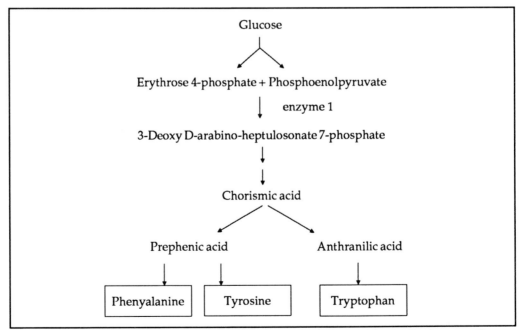

8.5.3 Evaluation of the production cost of L-phenylalanine by fermentation

The method

The biosynthesis of phenylalanine in organisms such as *E. coli* proceeds according to the following simplified scheme (many intermediates have been omitted - Figure 8.21).

Figure 8.21 Metabolic pathway leading to phenylalanine synthesis (see text for details of enzyme 1).

DAHP synthetase

In Figure 8.19, enzyme 1 is 3-deoxy D-arabinoseheptulosonic acid synthetase (called simply DAHP synthetase). DAHP synthetase is feedback inhibited by each of the end products of the pathway (that is Phe, Tyr and Trp).

SAQ 8.9	Which of the following auxotrophic mutant would be most likely to overproduce Phe?
	1) Phe⁻, 2) Tyr⁻, 3) Trp⁻, 4) Phe⁻ Trp⁻, 5) Phe⁻ Tyr⁻, 6) Trp⁻ Tyr⁻,7) Phe⁻ Trp⁻ Tyr⁻.

E.coli will grow in an ammonium-salts-glucose medium, the auxotrophic mutant will require additionally, the amino acids which it is incapable of synthesizing.

∏ Suppose the aim of a process is to manufacture 1000 tonnes (10^6 kg) of phenylalanine per year. Phenylalanine concentrations of 20 g l^{-1} can be expected and the yield of phenylalanine from glucose is 0.13g phenylalanine per g glucose consumed. Tyrosine is required at 0.21 g l^{-1} and tryptophan at 0.11 g l^{-1}. A reasonably-sized industrial scale fermenter might have a working volume of 150 000 l ($150m^3$). If the fermentation is operated batchwise, with a fermentation time of 72 hours (48 hours for the fermentation and 24 hours down time for preparing for the next fermentation run), how many fermenters will be required? (Try to work this out before reading on).

Production level is 20 g l^{-1} or 20 kg m^{-3}. Each batch of 72h thus produces 150 x 20 = 3000 kg phenylalanine. If we assume we could work the fermenter non-stop, we could run about 122 batches per year. In a year, each fermenter would produce 3000 x 122 = 366 000 kg of phenylalanine. Thus, to produce 10^6 kg would require 3 fermenters.

We are now going to do something quite different to the rest of this text. We are going to think about the economics of a process in detail. We will have to make some assumptions during this discussion so our final conclusions may not be very accurate. Nevertheless, if you follow the discussion carefully you will gain some insight into the cost of an industrial process.

Investment (capital) costs

∏ For a working volume of 150 m^3, the actual size of the fermenter would be about 200m^3. As a general rule of thumb, the investment (capital) cost of a fermentation unit is US $50 per l^{-1} actual size (1990 price). What would the investment cost of the system described above be?

The cost would be 200 000 x 3 x 50 = US $30 million, for three fermenters.

The investment costs include the cost of land, buildings, fermentation equipment, recovery equipment and all items required to make the plant operational.

Production costs

The production costs can be broken down into variable costs, direct costs, and plant gate costs. Let us examine these in turn.

Variable costs are the cost of raw materials used in the fermentation. Major costs for raw materials are those for glucose, tyrosine, tryptophan and mineral salts. The costs of these materials are given in Table 8.3 (Approximate 1990 prices).

Thus, the medium ingredients contribute US $4.46 per kg phenylalanine. Additional variable cost are the costs of other chemicals which may be used (for example in

	kg/kg Phe	price US $/kg	US $/kg phenylalanine
Glucose	7.5	0.40	3.0
Salts	0.53	0.1	0.53
Tyrosine	0.011	30.0	0.33
Tryptophan	0.006	100.0	0.6
		Total	**4.46 US $/kg**

Table 8.3 Raw material costs for phenylalanine fermentation.

downstream processing) and the cost of utilities (oil, gas, electricity and water). It is difficult to put an exact figure on these, as their cost varies from country to country. However, representative figures would be about 10% of total variable costs, so we will say US $0.50 per kg phenylalanine.

Production costs	(US $/kg phenylalanine)
a) Variable costs	
raw materials	4.46
utilities	0.50
Total	4.96
b) Direct costs	
direct labour	0.80
indirect labour	0.32
salaried payroll	0.68
associated payroll	0.54
maintenance salaries, laboratory etc	0.90
supplies and expenses for:	
- maintenance	0.90
- supplies	0.20
Total	4.34
c) Plant gate costs	
depreciation (15 years)	2.00
taxes and insurance	1.20
plant overhead	0.48
Total	3.68
Total production costs (fixed and variable) are thus:	**US $12.98/kg phenylalanine**

Table 8.4 Evaluation of production costs of L-phenylalanine fermentation (based on production of 1000 tonnes per year).

Direct costs are the cost of maintenance, supplies storage, labour and laboratory facilities. If we assume that 10 persons are needed to operate the plant in four shifts each day all the year through, this means a payroll of 40 people. At an average salary of US $20 000 per year, the total direct cost of labour would be US $800 000. Indirect labour costs account for 40% of direct labour and so a further US $320 000 must be added for this. The cost of salaried personnel (secretarial and managerial staff) is about 85% of direct labour = US $680 000 and a further 30% of the total payroll is used in associated payroll costs (for example contractual labour) = US $540 000. The cost of salaries of maintenance and laboratory staff are estimated at 3% of total investment costs (3% of US $30 million) = $900 000. The cost of these factors per kg phenylalanine are given in Table 8.4. The cost of supplies and expenses for maintenance and cost of other supplies is also given.

Plant gate cost are other costs, such as the depreciation in value of the plant (the plant will have been considered to have lost its value after a period of 15-25 years), taxes, insurance and other overheads. Depreciation of US $30 million over 15 years gives an annual depreciation cost of US $2 million.

Taxes and insurance will be in the order of 4% of the total capital cost (4% of US $30 million) = US $1.2 million. Overheads are estimated at about 60% of the total direct labour costs (60% of 800 000) = US $480 000.

Given these estimates of the production costs of phenylalanine, assessments of the likely profitability of the product can be made. Objective procedures to aid such assessment are based on the return of investment (ROI) as the criterion.

This is calculated from the following assumption (this is a business studies derived relationship, we will not explore the way in which it has been derived):

- ROI = [(A+9B)/10] + (C+D)

Where:

- ROI = Return of investment

- A = earnings after taxes (50%) in the first year

- B = net profit from operations in first year

- C = original fixed capital investment

- D = working capital (25% of net sales)

Let us assume that the investment costs must be recouped in ten years and that the selling price of L-phenylalanine is US $30-40 kg^{-1}. Thus, the sales value of the 1000 tonnes of product is US $30-40 million. About 30% of this sum will be accounted for by distributions and sales cost, so the net sales value will be about US $21-28 million.

Π How much profit will the 1000 tonnes of phenylalanine make? (Subtract the cost of producing the phenylalanine from the value of its sales).

With net sales of $21-28 million, and cost price of $13 kg^{-1} ($13 million for 1000 tonnes) the gross profit will be $8-15 million.

Assume that 12.5% of the net sales cost will have to be spent on administration and research and development (R & D) (US $3.0-3.5 million), so the profit will be of the order

US $5.0-11.5 million. On top of this, in the first year, start up costs amount to 10% of the capital costs (US $3 million), so in the first year the net profit will be US $2.0-8.5 million. Assuming taxes account for 50% of net profit, the actual earning will be US $1.0-4.25 million in the first year and in subsequent years, US $2.5-5.75 million.

Given the figures above, what will the ROI of this process be?

Using the equation given above and using the values:

- A = $1.0-4.25 million

- B = $2.0-8.5 million

- C = $30 million

- D = $5.25-7.0 million

Calculating first the lowest value:

- ROI = $[(1 \times 10^6 + 9 \times 2 \times 10^6)/10] + (30 \times 10^6 + 5.25 \times 10^6)$ -We can divide the equation through be 10^6.

- $= [(1 + 18) \div 10] + 35.25$

- $= [19 \div 10] + 35.25$

- $= 1.9 + 35.25$

- $= 0.05\ (5\%)$

Calculating the highest value:

- ROI = $[(4.25 \times 10^6 + 9 \times 8.5 \times 10^6) + 10] + (30 \times 10^6 + 7 \times 10^6)$

- $= [(4.25 + 76.5) \div 10] + 37$

- $= [80.75 \div 10] + 37$

- $= 8.075 \div 37$

- $= 0.22\ (22\%)$

You will appreciate that several important generalisations and assumptions have been made during these calculations. This was necessary because costs differ from country to country and much of the data on production processes such as this are not published (most companies regard such information as confidential). However this exercise will have given you an insight into the economics of amino acid fermentation. Let us now compare this to phenylalanine production by precursor addition.

8.5.4 Evaluation of the production cost of L-phenylalanine by precursor addition

The method

L-phenylalanine can be produced from the precursor phenylpyruvic acid, by addition to whole cells of an appropriate organism. The enzyme responsible is an aminotransferase and another amino acid (such as aspartic acid or glutamic acid) is used as amino donor. The process involves two stages: the first in which the cells are produced in culture and the second stage in which the precursor is added to the cells for conversion.

Step 1 - Cell production

Cell production is normally carried out using a fed-batch process. The carbon and energy source (such as glucose) are fed to the culture, so as to limit the growth rate and reduce the oxygen demand to a level in which high dissolved oxygen concentrations are maintained. After about 15 hours biomass concentrations of about 10 g l^{-1} can be reached.

Step 2 - Conversion

The precursor can be added directly to the culture during the stationary phase or the cells can be concentrated by centrifugation and resuspended at high concentration in another fermenter. The advantage of the latter method is that the cells can be suspended in a mineral salts, glucose maintenance medium (which does not allow growth) and so the concentration of contaminating by-products is reduced, and phenylalanine recovery is easier.

∏ Why do you think that resting (non-growing) cells are required for bioconversion?

If the cells were actively growing they could use the precursor and/or the amino acid product for growth (production of cellular proteins). By using stationary phase cells or cells in maintenance medium (in which lack of a nitrogen source prevents growth) then the precursor and amino acid product will not be used up for growth.

The conversion has been carried out using *E. coli* and *Pseudomonas fluorescens*. General reaction conditions are:

- 30-40°C;

- >50% dissolved oxygen tension (>50% of the saturation level);

- cell concentration 10-20 g l^{-1};

- substrate concentration 10-30 g l^{-1};

- amino donor (glutamic acid or aspartic acid) 35 g l^{-1}.

Under these conditions, conversions of nearly 100% can be achieved within 8h of batch operation.

When using free whole cells (in suspension) enzyme activity is lost within a few days (due to autolytic breakdown of the cells and the aminotransferase enzyme). Enzyme activity can be greatly stabilised if the cells are immobilised in alginate beads, which allows use for up to 60 days. However, to achieve high activity over such periods the cofactor pyridoxal phosphate has to be added to the reaction medium.

The cost

For fermentation of the amino acid, we have seen that the organisms are grown up in culture and the amino acid extracted. Precursor addition involves growing the organism up in fermenters in a similar way and then, in addition, feeding precursor to the organisms. Productivities of 3.5 g l^{-1}h^{-1} can be achieved throughout an 8h conversion period. Thus the system could produce 3 batches per day.

$\prod$ Using the data above, calculate the size of reactor required to produce 100 tonnes of phenylalanine per year.

With a productivity of 3.5 g l^{-1}h^{-1} (3.5 kg m^{-3} h^{-1}), this equals 3.5 x 24 = 84 kg m^{-3} day^{-1}. This equals 84 x 365 = 30 660 kg m^{-3} year^{-1}. Thus to produce 100 000 kg per year would require a reactor volume of 100 000/30 660 = approximately 3m^3.

Using these data, the costs for the production of L-phenylalanine by precursor addition from phenylpyruvic acid can be calculated (Table 8.5).

Production costs	(US $/kg phenylalanine)
Cultivation of organisms	3.28
3m^3 reactor	0.38
Product recovery	10.00
Labour	2.25
Utilities	0.59
Substrate (phenylpyruvic acid)	10.50
Amino-donor (L-aspartic acid)	8.05
Total	**US $35.05/kg phenylalanine**

Table 8.5 Estimated costs of production of L-phenylalanine by precursor addition (based on production of 100 tonnes per year).

$\prod$ Compare these costs with those of the fermentation route. Which looks like the cheapest?

At a cost of US $ 35.05 kg^{-1}, precursor addition is a more expensive method than fermentation (at a cost of $12.98 kg^{-1}).

8.5.5 Evaluation of the production cost of L-phenylalanine by enzymatic synthesis from trans-cinnamic acid

The method

L-phenylalanine can be produced from trans-cinnamic acid by the action of phenylalanine ammonia lyase. The enzyme can be obtained in high levels in the yeast *Rhodotorula*. The yeast is grown in medium containing L-isoleucine, as this stabilises the phenylalanine ammonia lyase throughout the growth period.

When used in suspension or immobilised, the whole cells of the yeast lose enzyme activity fairly rapidly. Reaction conditions are:

- 35°C;
- cell concentration 60-100 g l^{-1};
- substrate concentration 10-20 g l^{-1};
- ammonia concentration 90-100 g l^{-1};
- pH 10.

High ammonia concentration is necessary to prevent the occurrence of the reverse action of the enzyme. At a substrate concentration 10 g l^{-1}, a conversion level of 88% can be achieved in 15h batch operation. At substrate concentrations 20 g l^{-1}, conversion levels are reduced due to enzyme inhibition by phenylalanine.

The cost

With bioconversion from trans-cinnamic acid, productivities of 1 g l^{-1} h^{-1} (1 kg m^{-3} h^{-1}) are achieved. This equals 24 kg m^{-3} day^{-1} and 24 x 365 = 8760 kg m^{-3} $year^{-1}$ (assuming continual operation). Thus to produce 100 000 kg would require a reactor of volume 100 000/8760 = approximately 11 m^3.

From this data the cost of L-phenylalanine by bioconversion from trans-cinnamic acid can be calculated (Table 8.6).

Production costs	(US \$/kg phenylalanine)
Cultivation of organisms	6.13
11m^3 reactor	1.31
Product recovery	10.00
Labour	2.25
Utilities	0.97
Substrate	11.36
Total	**US \$32.02/kg phenylalanine**

Table 8.6 Estimated cost of production of L-phenylalanine by bioconversion of trans-cinnamic acid. The cost of ammonia is relatively low and has been ignored in the above estimation.

8.5.6 A comparison of L-phenylalanine production processes

productivity and reaction time

The process conditions for the three types of process we have considered are shown in Table 8.7. It is apparent that, in terms of productivity and reaction time, precursor addition is most favourable.

	Fermentation	Precursor addition	Bioconversion from trans-cinnamic acid
Productivity (g l^{-1} h^{-1})	0.6	3.5	1.0
Time to reach maximum levels of product (h)	24.0	8.0	15.0
pH	7.0	7.5	10.0
Temperature (°C)	35.0	35.0	35.0
Biomass concentration (g l^{-1})	20.0	10.0	70.0
Amino donor	-	L-amino acid	NH_3

Table 8.7 Process conditions for phenylalanine production by fermentation; bioconversion by precursor addition and bioconversion of trans-cinnamic acid.

comparability of direct costs

The distribution of costs in various categories for the three processes we have considered are shown in Figure 8.22. The direct costs (cost of labour, maintenance, laboratory facilities and so on) are roughly equal for each process. The variable costs (cost of raw materials and utilities) are greatest for the bioconversions, due mainly to the requirement for expensive substrates and amino donors. The overall cost of producing the organism, conversion and product recovery is also greatest for the bioconversions.

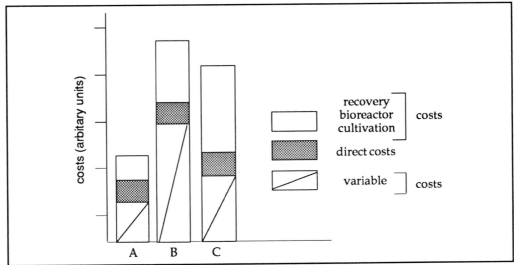

Figure 8.22 Comparative production costs for L-phenylalanine production by fermentation, A bioconversion by precursor addition, B and bioconversion from trans-cinnamic acid, C.

In summary, it would seem that L-phenylalanine production by fermentation is the most favourable process economically. However *E. coli* is generally not regarded as a suitable organism for the production of food additives and this could restrict application of such a product in the food industry. In general, fermentation systems can be operated at scales of 100 m³ or more, whereas bioconversion systems are more limited in scale (especially if immobilised cells are used). This means that in our economic comparison, the fermentation process benefitted from the economy of large scale.

It is possible that genetic engineering could be applied to improving bioconversions, by producing enzymes with greater activity and/or stability. This could make bioconversions more attractive economically.

Summary and objectives

This chapter has described the production and uses of amino acids in the food industry. It has examined a variety of strategies for producing amino acids on a large scale. These include both the use of natural and manipulated micro-organisms and of single, purified enzymes. The text has covered issues of production and product recovery and has made comparisons between different strategies. Now that you have completed this chapter you should be able to:

- outline the uses of amino acids in the food industry;

- describe the uses of wild-type organisms, auxotrophic mutants and regulatory mutants in amino acid fermentation;

- outline the application of genetic engineering to amino acid fermentation;

- describe the technology of fermentation and product recovery in amino acid fermentation;

- outline problems associated with amino acid production by fermentation;

- describe the application of resting whole cells to amino acid production;

- describe the application of enzymes to amino acid production;

- compare the economics of the production of phenylalanine by fermentation, precursor addition and enzymatic synthesis.

Responses to SAQs

Responses to Chapter 2 SAQs

2.1   1) False. α-Amylase does not have application at the mixing stage.

2) True.

3) False. α-Amylase increases loaf volume by decreasing amylose concentrations.

4) True.

2.2   The factors favouring enzyme use are 1), 2) and 3). 4) produces equivalent activity to chemical processes. Chemical processes can be operated on dry reactants 5), and solids 6) (which melt at the high reaction temperatures used).

2.3   This is a little bit of a trick question which needs to be read carefully. The factors contributing to protection of eggs against spoilage by Gram-negative bacteria are 3) and 4). 1) refers to lysozyme which is only lytic for Gram-positive bacteria. Immunoglobulins, 2) are present in milk, but not eggs. Although eggs lack catalase, 5), this is not involved in protection against spoilage here. Bacteriocins, 6), are not found in eggs.

2.4   1) True.

2) False. H_2O_2 is itself a good preservative. Its removal does not in its own right aid preservation.

3) False. Lactoperoxidase requires H_2O_2 for its preservative action. Removal of H_2O_2 decreases the effect of lactoperoxidase.

4) False. H_2O_2 is the product of xanthine oxidase activity, its removal decreases the preservative effect.

2.5

	A		B
i)	Butter	b)	Diacetyl
ii)	Yoghurt	c)	Acetaldehyde
iii)	Roasted	f)	2-Methoxyl 3-isobutylpyrazine
iv)	Fruity	e)	Ethylacetate (an ester)
v)	Peach	d)	α-Decalactone
vi)	Blue cheese	a)	2-Pentanone (this is a methylketone)

2.6 This is quite a difficult question to answer because these organisms are mentioned in several places in the text. The organism producing the largest number of the flavour compounds listed is d) - *Lactococcus lactis*. This organism (subsp. *lactis* var. *diacetyl lactis*) produces diacetyl. It also produces esters and pyrazines. Although not specified in the text it also produces acetaldehyde. If you got this question right - well done! You have read the text very carefully.

2.7

Food/Food Product	Proteases	Amylases	Lipases
Bread	3	3	-
Tenderised meat	3	-	-
Cheese	3	-	-
Interesterified oil	-	-	3
Invert sugar	-	-	-
Modified whey	-	-	-
Modified protein thickener	3	-	-
Organic acids	-	-	-
EMC	3	-	3

Responses to Chapter 3 SAQs

3.1   The correct matching is as follows:

lacY permease	*Lb. Acidophilus*	galactose utilised
PTS	*L. lactis*	PEP-dependent
lacS permease	*Lb. bulgaricus*	galactose excreted

3.2   Your figure should have looked something like this

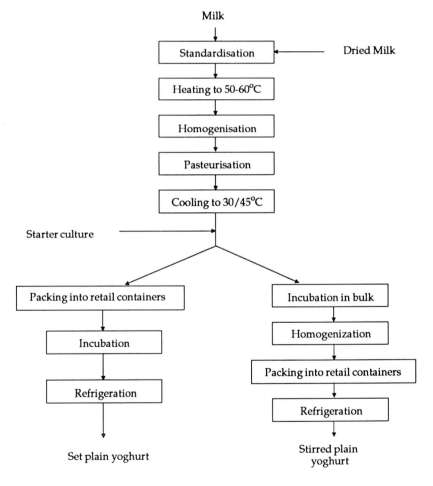

3.3   The correct response is 5). Protease activity would be needed to produce the amino acids which are formed in yoghurt manufacture by *Lb. bulgaricus* and which *St. thermophilus* requires.

3.4   Completions 2) and 3) are true. 1) is untrue as the anaerobic conditions maintained in sauerkraut prevent spoilage by (aerobic) filamentous fungi. However if the sauerkraut was stored in aerobic conditions 1) would be true.

3.5   Completions 1), 2) and 4) are true. 3) is not normally true: although lactic acid does eventually inhibit the growth of lactic acid bacteria, the enzymes which they produce do not normally lead significantly to spoilage.

3.6   The correct matching is as follows.

yoghurt	*Streptococcus thermophilus*	neither
sauerkraut	*Leuconostoc mesenteroides*	low water activity
cheddar cheese	*Lactococcus lactis*	both
dry sausage	*Pediococcus acidilactici*	both

Yoghurt is not dehydrated or salted. Sauerkraut is salted. Cheese and sausage are both salted and dehydrated.

3.7   Your flow diagram should have looked something like this:-

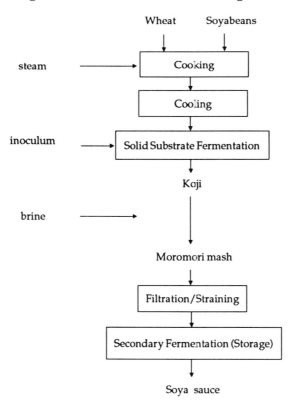

3.8   The correct matching is as follows.

butter	citrate utilisation	*L. lactis*
yoghurt	threonine degradation	*Lb. bulgaricus*
dry sausage	nitrate reduction	*Micrococcus spp.*
koji	starch degradation	*Aspergillus oryzae*
soya sauce	alcohol production	*Saccharomyces rouxii*
tempeh	cellulose degradation	*Rhizopus oligospora*

Responses to Chapter 4 SAQs

4.1   The correct terms are 3) and 4). Clearly the culture does not contain a single strain nor is it a monoculture therefore 1) and 8) are incorrect. Multiple strain starters 2), are starters which contain up to about 6 strains therefore this is incorrect. Likewise D- and DL- starters contain lactococcal strains as aromabacteria (5 and 6 are incorrect) O starters 7) contain no aromabacteria.

4.2

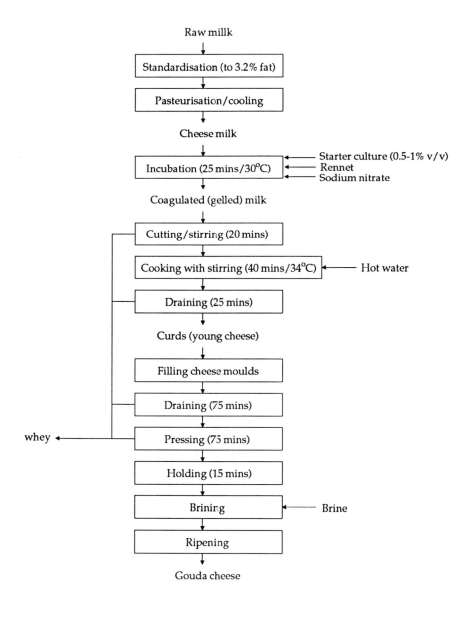

Raw millk
↓
Standardisation (to 3.2% fat)
↓
Pasteurisation/cooling
↓
Cheese milk
↓
Incubation (25 mins/30°C) ← Starter culture (0.5-1% v/v)
                          ← Rennet
                          ← Sodium nitrate
↓
Coagulated (gelled) milk
↓
Cutting/stirring (20 mins)
↓
Cooking with stirring (40 mins/34°C) ← Hot water
↓
Draining (25 mins)
↓
Curds (young cheese)
↓
Filling cheese moulds
↓
Draining (75 mins)
↓
whey ← Pressing (75 mins)
↓
Holding (15 mins)
↓
Brining ← Brine
↓
Ripening
↓
Gouda cheese

4.3   1) The correct term is b). The mechanism of incorporating modified plasmid DNA is transformation. Conjugation involves transfer of plasmids by cell-cell contact and transduction involves transfer by phage. Transduction is a natural gene transfer process by which genetic material is packaged into the head of a bacteriophage and transferred during the subsequent infection process. Protoplast fusion is genetic recombination after fusion of protoplasts to produce a diploid cell. Recombination of DNA could occur as a result of each of the gene-transfer systems listed.

2) The answer is homologous. The process described only involves genetic information from a single strain. Such a modification would probably be readily accepted for use in cheese manufacture provided it passed all other relevant criteria.

4.4   The correct completion is 1) and 3). An O-starter lacks aromabacteria. An L-starter contains *Leuconostoc spp*. Removing the aromabacteria from the L-starter would thus produce the O-starter.

4.5   The key to getting the correct answer here is to recognise that the question is aimed at mature (ie aged) cheese.

The correct completion is 3). 1) is incorrect; rennet is added to clot milk (see Chapters 3 and 5) and coagulation does not affect directly the ripening process. 2) may also be correct in itself but does not affect directly the ripening process. 4) is incorrect, as it is a deficiency (lack of) proteinase enzymes which characterises Prt⁻ strains.

4.6   The correct completion is 1). Prt⁻ organisms rely on Prt⁺ organisms to produce proteinases and so supply amino acids. If all Prt⁺ organisms disappeared from a culture, the result would be a shortage of amino acids on which Prt⁻ organisms could grow.

4.7   The correct matching is shown below:

i)	New Zealand system (before 1970)	c)	Paired strains (phage resistant)	3)	Rotation
ii)	Whey-derived system	b)	Single strain	2)	No rotation (derive resistant strains upon phage infection
iii)	Multiple-strain starter system	a)	3-4 mixed strains	1)	No rotation (no action taken upon phage infection)

4.8   Correct choices are 4) and 5). Traditional mixed strain starters have a wide mixture of strains within them. Although plasmids will be lost, the inability to ferment lactose (Lac⁻) or produce proteinase (Prt⁻), will not predominate, as such strains will not be able to grow better in milk than Lac⁺ or Prt⁺ strains (see SAQ 4.6). Thus points 1) and 3) are incorrect. 2) is incorrect as mixed strain starters are the easiest of all to maintain and produce, since this only involves simple subculture.

4.9   The correct responses are as follows:

1) + Calcium ions promote phage attachment and infection;

2) - This would ensure the maintenance of phage insensitivity by killing off any phage-sensitive mutants;

3) + The phage-sensitive strain would probably suffer phage infection. This could well produce mutant or recombinant phages to which the starter was sensitive;

4) - This would help to confer phage insensitivity on the starter;

5) Plasmids in *Lactococcus spp.* can carry genes that confer phage insensitivity (Table 4.3.) Their loss would make the organism phage-sensitive. Loss of Prt^+ or Lac^+ genes might lower the growth rate and so promote phage-insensitivity (see SAQ 4.8). Therefore you could have scored this part as +, -, or ±, as all are potentially correct;

6) + See Section 4.5.1 for a reminder of the reasons.

4.10   Why change a winning formula! The best option is 1), the lack of failure indicates that the starter you are using is insensitive to the phage(s) present in the production unit. P-starters are infected by high levels of phages, and their use in the unit would contaminate it. Thus, when you subsequently use your single-strain starter it could suffer infection from one of the contaminating phages from the P-starter. Whey-derived, or BIM starters or some organisms in P-starters might not be insensitive to the phages present in the unit. Their use would cause multiplication of existing phages and so allow possible formation of recombinants or mutants to which your single-strain starter is sensitive.

4.11   a) exerts an effect on stage v) in Gouda and stages iv) and v) in Cheddar - Salt is not added until these stages in cheese production.

b) exerts an effect on stage v) - Lactose becomes fully used when ripening starts.

c) exerts an effect on stage i) - During the cheese-milk fermentation oxygen must be used up to ensure fermentation. In all other stages anaerobic conditions prevail.

d) exerts an effect on all Stages - At any stage the level of any activity will be proportional to the number of organisms present.

4.12   You should have matched the following:

i) with d and stage 3 - Temperature-insensitive mutants ferment lactose at the high cooking temperature;

ii) with c and stages 1/2/5 - The Prt mutants lack proteinases. Proteinases contribute to fermentation and ripening;

iii) b and all stages - Carbon dioxide production continues throughout the stages of manufacture. It is probably most rapid during stages 1 and 2, when lactose concentrations are highest;

iv) with a and all stages - It is important that phage-insensitivity is maintained throughout the whole of the cheese production process.

4.13   The correct sequence is as below.

Medium sterilisation.

Inoculation with stock culture.

Incubation (20°C 20h).

Centrifugation.

Cooling to 1°C.

Packaging.

Freezing to -35°C.

4.14   1) True.

2) False. Peptidases do not form peptides from proteins. Peptidases degrade peptides. If the peptides are responsible for bitter flavour, then peptidases will be beneficial.

3) True.

4) True.

Responses to Chapter 5 SAQs

5.1　1) is false as acidification alone does clot milk, as in the production of acid cheeses and yoghurt. 2) is true. 3) is false as lactic acid bacteria do not produce clotting enzymes.

5.2　1) and 3) are true. Rennet from younger calves contains a higher proportion of chymosin. Non-specific proteolysis by pepsin weakness the gel network. 2) is false as non-specific proteolysis is not promoted by high chymosin levels. 4) is false because specific cleavage of κ-casein during coagulation is about the same for bovine and half rennet.

5.3　You will have to forgive us for including in this book at least one 'tricky' question! You will probably have answered 4). In fact all the enzymes listed COULD produce such fragments. The enzyme listed in 4) produces such fragments preferentially. However both chymosin and the protease from *M. miehei* will produce a small proportion of such fragments. This is because they preferentially cleave the κ-casein between Phe 105 - Met 106. The pepsin in rennet will produce such fragments at random by non specific cleavage. If you did answer this correctly award yourself a gold star!

5.4　For coagulation you should answer 1), because specific hydrolysis of κ-casein by rennet enzymes is the essential primary step.

For ripening you should answer 3) and 4) as the most important enzymes. Enzymes 1), 2) and 5) may also contribute but to a much lesser extent.

5.5

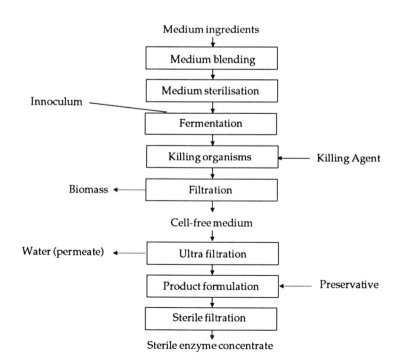

5.6   1) GMMO = genetically modified micro-organism, GMO = genetically modified organism.

2) Mutagenesis, self-cloning of some non-pathogenic organisms (see Appendix 1, Section 1.4).

3) Group II - *Salmonella typhi* is a pathogen and does not fall into group I (see Appendix 1, Section 1.5).

4) Type B operations are large scale (10 litres or greater) operations used for commercial purposes (Appendix 1, Section 1.5).

5) To the competent authority of the Member State within whose territory the release is to take place (Appendix 1, Section 2.4). This notification should contain information set out in 5 headings.

General information.

Information relating to the GMO.

Information relating to the conditions of release and receiving environment.

Information relating to the GMO and the environment.

Information on monitoring, control, waste treatment and emergency response plans (Appendix 1, Section 2.4).

6) The competent authority of the Member State where they are placed on the market for the first time (Appendix 1, Section 2.5.2).

5.7   1) Research or Small Scale Guidelines, Large Scale Guidelines and Regulations, Relevant Product Regulation (see the beginning of Appendix 2).

2) *K. lactis* has been given the lowest risk category VMT. In France the lowest risk category is designated L_1L_1. Note that in the Uk this same category is designated Group I.

3) The three basic principles which should be met before being placed on the market are:

• the food must not be dangerous to public health;

• the food must not impede fair trade;

• the food shall not mislead the consumer (see page 2 of Appendix 2).

4) You could have made an extensive list. In Appendix 2 we suggested the following section:

Product organism (pathogenicity, stability of DNA, characterisation of DNA);

Product toxicology (physico-chemical characterisation, microbiological specifications, chemical specifications);

Product toxicology (short term studies, sub-chronic studies, mutagenicity studies, allergenicity studies);

Product efficiency.

You may well have listed more specific tests to be included under each of these headings.

5.8   The correct completion is 2).

3) and 4) enable the problem mentioned in 2) to be overcome, but you do not develop a product simply because the technology to do so exists, you only develop a product if enough cannot already be made or if the new product is better or cheaper than the existing one. In some respects recombinant chymosin is better than rennet but the cheese industry regards it as a rennet substitute (an equivalent rather than a better product), so 1) is not correct.

5.9   The likely sequence would be:

1) Select host organism

2) Introduce heterologous DNA into host organism.

3) Identify product of heterologous gene in host organism.

4) Develop a fermentation medium.

5) Develop a method of killing the organism in the fermentation medium.

6) Scale up the fermentation process.

7) Conduct manufacturing trials.

8) Commercial operation.

The first stage is to produce a recombinant organism 1) and 2) and then ensure it is producing the required product 3). Only then would an (optimalised) medium to grow the recombinant organism and stimulate product be developed 4). The killing method for use in this medium can then be developed 5). Scale up 6) would follow medium and process development. At around the same time trials might be performed 7) to see how the product works in practice. This would be done as soon as enough product could be produced. Commercial operation is the final stage.

Responses to Chapter 6 SAQs

6.1
1) The answer is amylopectin not starch. Starch is a mixture of two types of polymers (amylose and amylopectin).

2) DE is a measure of the amount of reducing sugar in a syrup. When starch is hydrolysed, reducing sugars are exposed at the site of hydrolysis. Consequently the DE value will rise.

3) Saccharification is the hydrolysis of starch, so during saccharification the DP (degree of polymerisation) value will fall.

4) It is called gelatinisation - the other terms are incorrect.

5) True - laevulose is another name for fructose which would be the product of the complete hydrolysis of a fructose polymer.

6.2

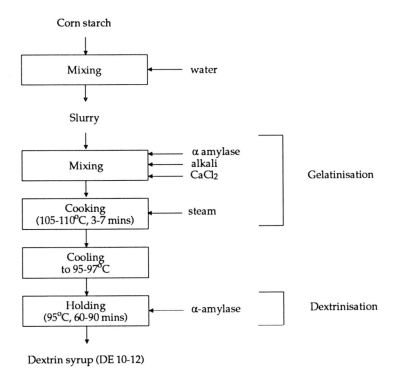

6.3
The correct answer is 4). Amylose is unbranched and thus unaffected by pullulanase. Amyloglucosidase does not act rapidly on amylose as it acts only on non-reducing ends. α-Amylase on its own, does not produce significant quantities of glucose. Amyloglucosidase and α-amylase in combination, 4) produce glucose most effectively. You could also answer 6). This would be as effective as 4), but the pullulanase included would be wasted.

6.4 The correct answer is 6). Pullulanase would break the α-1,6 bonds in amylopectin, liberating amylose which would be cleaved by α-amylase and the resulting dextrins would be rapidly hydrolysed to glucose by amyloglucosidase.

6.5 Each enzyme produces some glucose from starch overnight at 60°C. Both α-amylase from *B.licheniformis* and amyloglucosidase would do this. However, amyloglucosidase is not active at 110°C, whereas α-amylase is.

Thus, enzyme A is amyloglucosidase and B is α-amylase.

6.6

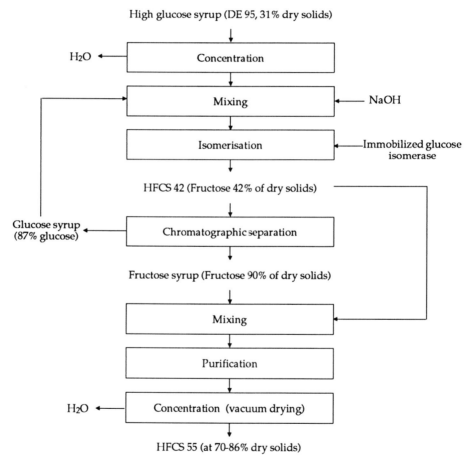

6.7
1) False - Low A_w is used for preservation purposes. In ice cream the low temperature does this.

2) True - This means it is easier to make soft ice cream.

3) False - It is used to provide the vegetable with a crisp texture.

4) False - The high baking temperatures give bakery products a crust. Sugars contribute to colour by caramelisation.

6.8   1) The major production cost in HFCS is the cost of starch US $8.80 per 100 kg.

2) The value of the by-products from wet milling reduces the production cost of HFCS by about 30% (without the reduction of US $11.0, the total cost would be 23.85 + 11.0 = 34.85.

3) The greatest cost of an enzyme is the cost of glucose isomerase US $0.38 per 100 kg HFCS).

4) The cost of all the enzymes used in HFCS production contributes about 5% to the total production cost (0.38 + 0.17 + 0.35 = 0.90 in 23.85 = about 5%).

6.9   1) False - *E. coli* does not produce α-amylase. Cloning vectors from *E. coli* have been used.

2) True.

3) False - Although *B. stearothermophilus* produces α-amylase, production levels are low.

4) True

6.10   1) False - Organisms with multiple gene copies can also be produced by incorporation of multiple copies of plasmids into the cells.

2) True.

3) False - see 2).

4) True.

6.11   The factor favouring the production of HFCS rather than invert sugar is 1). Factors 2) and 3) also apply to invert sugar. 4) does not favour HFCS against invert sugar, as invert sugar requires only one enzyme for its formation.

6.12

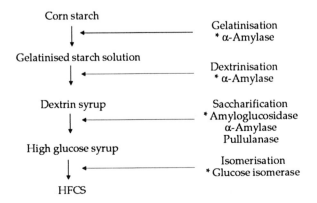

Responses to Chapter 7 SAQs

7.1   The best substrate for endo-PG would be 3) - pectic acid. This is because the enzyme only cleaves bonds adjacent to molecules which are non-esterified and pectic acid has the greatest proportion of these.

7.2   Endo-polygalacturonase. This is the only enzyme that has the activity described.

7.3   2) is the correct answer. PAL splits the glycosidic linkages next to sugar units having free carboxyl groups. PE removes methyl esters therefore producing a more suitable substrate for PAL. The converse is true for mixture 1) PE and PL. 3) PL and PAL would not interact in a mixture in any significant way. Our answer assumes that the pH would allow activity in each enzyme present.

7.4   The correct response is 5). Only PG would produce the given effects on pectin. Cellobiase is present as glucose is produced from cellobiose. Arabinosidases are not present, ie 3) and 4), as reducing sugars are not produced from arabinan. β-glucanases are not present, ie 1), 2) and 6) as reducing sugars are not produced from CMC.

7.5   The correct responses are as follows.

1) Significant amounts of pectin can only be degraded from the wall fragments by PL or PG + PE.

2) Cellulases are more effective than hemicellulases when each is used in isolation.

3) Cellulases are more effective than hemicellulases when each is used in combination with a pectinase.

4) The effect is least when cellulase or hemicellulase is added to PG.

5) The maximum effect occurs with the addition of cellulase to PG + PE.

6) The effect of adding cellulase to PL or PG + PE is synergistic (as the combined effect is greater than the sum of each of the individual effects - for example PL produces 12% sugar and CE 10%, yet in combination they release not 22% but 37%).

7.6   The correct answer is 3). 1) PAL would not be active in the low pH of fruit juice. 2) PG would not be active on high-methoxyl pectin on its own. 3) PE + DE would degrade both high- and low-methoxyl pectin. 4) PL would not be active on low-methoxyl present, so if the pectin was of this type it would not be degraded. For the same reason 5) PL and cellulase and 6) PL and hemicellulase might not be effective.

7.7   The correct response is 3). PE demethylates polygalactopyranosyl chains. These demethylated regions complex Ca^{2+} ions and bind chains together. If pectin was degraded by PG or PL, the demethlylated regions are too small to form a complex with Ca^{2+} (see 7.3.1).

7.8   Cause 3) - arabinosidase B. This enzyme debranches arabinans and thus forms the linear chains which cause haze.

Prevent 4) - the mixture of endo-arabinase and arabinosidase A would be most effective in removing linear chains, so preventing haze.

Responses to Chapter 8 SAQs

8.1

1) True - Wild-type organisms do not generally excrete amino acids (Section 8.3.1).

2) False - Auxotrophic mutants can suffer from feedback inhibition in the same way as wild-type organisms.

3) True.

4) True.

8.2

1) The mutant most likely to overproduce H would be e) mutant 5, E⁻ J⁻. This organism could not form E or J, so the intermediates B and D would be shunted towards the production of H.

2) Mutant 5 would require the addition of E and J to the production medium.

3) If E was present at high concentrations it would inhibit enzyme 1, which would depress formation of intermediate B and so also depress formation of H. If J was present at high levels it would depress formation of G from D. Thus more D would be available to produce H (ie H production might be increased).

8.3

Production of amino acid by fed-batch fermentation is supported by 1), 4) and 5). Factors 2) and 3) are common to both fed-batch and batch cultures.

8.4

1) False - Broths are sometimes concentrated to improve the crystallisation process, but crystallisation is brought about mainly by adjustments of pH and temperature.

2) False - The reverse is true.

3) True - The biomass would clog equipment and interfere in other ways with product recovery.

4) True.

8.5

1) False - *P. dacunhae* is used in the closed-column reactor, not *E. coli*.

2) True.

3) False - Ammonia is required in the first stage using *E. coli*, not *P. dacunhae*.

4) True.

8.6

1) False - Aminoacylases resolve D,L-N-acetyl amino acids, they do not resolve D,L amino acids.

2) False - This enzyme racemises. α-amino-ε-coprolactam.

3) False - Ammonia lyases can use ammonia as amino donor.

4) False - The enzyme produces L-alanine from L-aspartic acid. In theory the enzyme can work in the opposite direction but the equilibrium constantly favours L-alanine formation under normal incubation conditions.

8.7   Biotransformation using immobilised enzymes rather than free enzymes is supported by 1), 3) and 4). Item 2) is common to both.

8.8   The correct completion is 2). The other statements can be true for precursor addition and other bioconversions.

8.9   The mutant most likely to overproduce phenylalanine is 6) Trp⁻, Tyr⁻. This mutant would not produce tryptophan nor tyrosine which would not feedback inhibit DAHP synthetase. Chorismic acid would therefore be solely directed towards synthesis of phenylalanine.

Appendix 1

1 EC directive on the contained use of genetically modified micro-organisms

The EC directive on the contained use of genetically modified micro-organisms is based on the OECD guidelines and is published in the Official Journal of the European Communities, L117, Volume 33, 8 May 1990.

1.1 Purpose of the directive

Article 1 gives the purpose of the directive as:

"to lay down common measures for the contained use of genetically modified micro-organisms with a view to protecting human health and the environment".

1.2 Scope

The scope of every regulation depends on the definitions and the exemptions.

This directive covers contained use of genetically modified micro-organisms. In the following paragraphs, therefore definitions of the terms genetically modified micro-organisms and contained use are given, and exemptions are explained.

1.3 Definitions

1.3.1 Micro-organisms

Micro-organisms are defined as:

"any microbiological entity, cellular or non-cellular, capable of replication or transferring genetic material".

It should be noted that this directive only covers micro-organisms, and not all organisms. It was argued that including all organisms in this directive would give an unacceptable delay in the implementation of the proposals. The Commission undertook to keep the whole biotechnology sector under review and make appropriate proposals to extend the scope of this directive to genetically modified organisms. The national Member States may maintain and adopt national measures for the contained use of organisms other than micro-organisms.

Genetically modified micro-organisms are in this document abbreviated as GMMOs.

1.3.2 Genetically modified micro-organisms

Genetically modified micro-organisms are defined as:

"micro-organisms in which the genetic material has been altered in a way that does not occur naturally by mating or by recombination".

In order to specify this, a list of techniques is given by which genetic modification can occur.

This non-limitative list is Annex Ia, of the Directive and contains:

- recombinant DNA techniques using vector systems;

- techniques involving the direct introduction of heritable material prepared outside the micro-organism, such as micro-injection;

- cell fusion and hybridisation techniques by means or methods that do not occur naturally.

It should be noted here that the scope of the regelation is not limited to recombinant DNA techniques. This gives recognition to the fact that a variety of molecular genetic transformation techniques, including recombinant DNA, are widely used and may have similar safety considerations.

1.3.3 Contained use

Contained use means:

"any operation in which micro-organisms are genetically modified or in which such organisms are cultured, stored, used, transported, destroyed or disposed of and for which physical barriers together with chemical and/or biological barriers, are used to limit their contact with the general population and the environment".

The key term here is the physical barrier.

1.4 Exemptions

The directive does not apply where genetic modification is obtained through the use of certain techniques (Article 3). These techniques are:

- mutagenesis;

- construction and use of somatic animal hybridoma cells;

- cell fusion of plants which can also be produced by traditional breeding methods;

- self cloning of certain non-pathogenic naturally occurring micro-organisms.

The directive does furthermore not apply to:

- the transport of GMMOs;

- GMMOs which have been placed on the market under Community legislation (Article 5).

1.5 System of the directive

1.5.1 Group I and Group II organisms/Type A Type B operations

Group I and Group II organisms

For the purpose of the directive, micro-organisms are classified in two groups (Article 4).

Group I - those satisfying certain criteria.

Group II - those other than group I.

Group I organisms have a long record of safe use and are considered to be safe when used under specific conditions. The criteria for Group I are given in Annex II of the directive, which gives criteria for the recipient and parental organisms (non pathogenic etc), to the vector and the insert used, and to the final GMMO. This Annex is based on the criteria for GILSP (Good Industrial Large Scale Practice) set up by the OECD.

For GMMOs of Group I, principles for good microbiological practice and of good occupational safety and hygiene shall apply. These principles are also based on the OECD report of 1986.

In addition to these principles, certain containment measures set out in an Annex shall be applied to ensure a high level of safety for GMMOs of Group II (Article 6).

Type A and Type B operations

For the purpose of the directive, a distinction is made between Type A and Type B operations.

Type A operations are operations used for teaching, research and development, or non-industrial or non-commercial purposes and which are of a small scale (eg 10 litres volume or less).

Type B operations are operations other than operations of type A.

1.5.2 System of the directive

The regulatory system of this directive contains two sorts of procedures:

- activities for which a notification is required;

- activities for which an authorisation is required.

The two distinctions can be combined to four possible activities:

- Type A operations with Group I - IA operations;

- Type B operations with Group I - IB operations;

- Type A operations with Group II - IIA operations;

- Type B operations with Group II- IIB operations.

In addition to this, the first use of an installation for an operation involving GMMOs is considered to be an activity for which a procedure is required.

Articles 6, 7 and 8 of the Directive assign specific procedures to each of the possible activities mentioned above and form the basis of the procedures.

The possible procedures are:

- to keep records of the work carried out and make them available to the competent authority (Article 9.1);

- a notification within a reasonable period before commencing the use (Article 8);

- a notification and a waiting period (Article 9.2 and 10.1);

- an authorisation (Article 10.2).

The possibilities are presented below

Type of operation	Procedure	Article
First use in an installation	notification within reasonable period in advance	8
IA operations	keep records	9.1
IB operations	notification and waiting period of 60 days plus additional 60 days at request of competent authority	9.2
IIA operations	notification and waiting period of 60 days plus additional 60 days at request of competent authority	10.1

1.6 Additional provisions

Articles 11 to 14 lay down some specific obligations for the Member States:

- to ensure that, where necessary, before an operation commences an emergency plan is drawn up and information on safety measures is available (Article 14);

- to ensure that, in case of an accident, the proper measures and steps will be taken (Article 15);

- consult, when necessary, with other Member States and inform the Commission (Article 16);

- to ensure that inspections are carried out (Article 17a).

1.7 Confidentiality

Article 19a:

"The Commission and the competent authorities shall not divulge to third parties, any confidential information notified or exchanged under this directive and shall protect the intellectual property rights relating to the data received".

The notifier indicates what information needs to be kept confidential. However, it is the competent authority which decides, after consultation with the notifier, which information shall be kept confidential.

2 EC Directive on the deliberate release of genetically modified micro-organisms

This directive of 23rd April 1990 is published in the Official Journal of the European Communities, L117, Vol 33, 8 May 1990.

2.1 Parts of the Directive

This directive consists of four parts:

Part A - General provisions;

Part B - Research and Development (R & D) and introductions into the environment other than placing on the market;

Part C - Placing products on market ;

Part D - Final provisions.

2.2 Part A: General provisions

2.2.1 Purpose of the Directive

The purpose of this directive is laid down in Article 1:

"to approximate the laws, regulations and administrative provisions of the Member State and to protect human health and the environment when carrying out a deliberate release or placing on the market of genetically modified micro-organisms".

In order to gain a better understanding of the purpose and background of this directive, the considerations in the preamble should also be studied.

In addition to the purpose of this directive Article 4 emphasises in general terms the obligations of Member States in accomplishing this purpose.

2.2.2 Scope of the Directive-definitions

Repeating what was explained earlier: the scope of every regulation depends on the definitions and exemptions.

This directive covers the deliberate release of genetically modified organisms. In the following paragraphs the definitions of the directive are explained. These definitions are summed up in Article 2.

Organism

Organism is defined in this directive as:

"any biological entity, capable of replication or of transferring genetic material".

Since the term 'biological entity' is open for multiple interpretation, an explanation is given in the statements for inclusion in the Council's minutes:

"This definition covers: micro-organisms, including viruses and viroids; plants and animals; including ova, seeds, pollen, cell cultures and tissue cultures from plants and animals".

Hereafter, the term genetically modified organisms is abbreviated to GMO.

Genetically modified organism

The definition of a genetically modified organism is analogous to the definition of a genetically modified micro-organisms, provided that the term 'micro-organism' is replaced by the term 'organism'.

Deliberate release

Deliberate release is defined in paragraph 3 of Article 2 as:

"any intentional introduction into the environment of a GMO or a combination of GMOs without provisions for containment such as physical barriers or a combination of physical barriers together with chemical and/or biological barriers used to limit their contact with the general population and the environment".

A further clarification is given in the Council's Statements:

"the introduction by whatever means, directly or indirectly, by using, storing, disposing, or making available to a third party".

By using the terms "without provisions for containment such as..." as a cross reference to the directive for contained use, a complementary system is achieved.

In other words: every activity that is not a contained use is regarded as a deliberate release.

2.2.3 Exemptions

The exemptions of the scope of this directive are laid down in Article 3:

"This directive shall not apply to organisms obtained through the techniques of genetic modification listed in Annex Ib". These include:

- mutagenesis;

- cell fusion of plant cells when the plant can also be produced by traditional methods.

The background of these exemptions is that these specific applications have been used in a number of applications and have a long safety record.

2.3 System

The system of this directive is based on two notions:

- the release of a GMO into the environment can have adverse effects on the environment which may be irreversible;

- GMOs, as well as other organisms, are not stopped by national frontiers.

These two notions led to the choice of a system whereby:

- every introduction of a GMO into the environment is subject to an authorisation by the competent authority of the country where the introduction takes place;

- before an authorisation is given, the competent authority consults the other Member States of the Community.

In addition to this, a distinction between Research and Development (R & D) and placing on the market is made.

2.3.1 R & D and placing on the market

In this directive, a distinction is made between:

- Research and Development (R & D) and other introductions into the environment than placing on the market (part B of the directive) ;

- Placing on the market (part C of the directive).

The result of this distinction and the system of the directive is that placing on the market involves a system of international consultation whereby the competent authority cannot take a decision without the agreement of the other Member States.

All other introductions into the environment (Part B, which basically consists of R & D introductions) are subject to an authorisation of the competent authority which may give its decision without the approval of other Member States, though be it that these introductions are also notified to the other Member States who may give comments.

The reason for this distinction, is found in the numbers and the spread of a GMO connected with placing on the market. When a product containing GMOs or consisting of a GMO is placed on the market, it will be spread all over Europe in, possibly, vast numbers under uncontrolled circumstances. Whereas R & D introductions are normally small scale introductions of a limited number of GMOs and under controlled circumstances.

2.4 Part B: Research and Development (R & D) and introductions into the environment other than placing on the market

The basis of part B of this Directive is laid down in the combination of the Articles 5, paragraph 1 and Article 6, paragraph 4.

Article 5, paragraph 1 states:

"Any person before undertaking a deliberate release of a GMO for the purpose of research and development or for any other purpose than placing on the market, must submit a notification to the competent authority of the Member State within whose territory the release is to take place".

Article 6, paragraph 4, states:

"The notifier may proceed with the release only when he has received the written consent of the competent authority, in conformity with any conditions required in this consent".

2.4.1 The notification

Article 5, paragraph 2, gives the requirements of a notification under part B of the directive:

"The notification shall include the information specified in Annex II".

Annex II is an indicative list of points of information set out under 5 headings:

- I   General information;

- II   Information related to the GMO;

- III   Information relating to the conditions of release and the receiving environment;

- IV   Information relating to the GMO and the environment;

- V   Information on monitoring, control, waste treatment and emergency response plans.

This Annex II is based on the OECD report of 1986. It is essential to realise that this Annex contains an indicative list, and that not all the points included will apply to every case.

2.4.2 Authorisation

The authorisation procedure is laid down in Article 6.

Paragraph 1: "On receipt and after acknowledgement of the notification the competent authority shall examine the conformity of the notification with the requirements of this directive".

Paragraph 2: "The competent authority, having considered where appropriate, any comments by other Member States, shall respond in writing to the notifier within 90 days by indicating either:

- that the release may proceed;

- that the release does not fulfil the conditions of this directive and the notification is therefore rejected.

For calculating the waiting period of 90 days, the period needed for the notifier to supply further information and the period in which a public inquiry is carried out, shall not be taken into account.

The third paragraph of Article 6 gives the steps to be taken when new information becomes available with regard to the risk of the product. In that case the notifier shall revise the information, inform the competent authority and take the necessary measures to protect human health and the environment.

2.4.3 International consultation

Within 30 days after the receipt of a notification, the competent authority shall send to the Commission a summary of the notification. The Commission shall immediately forward these summaries to the other Member States which may, within 30 days, present observations. It should be stressed here that these observations are not binding to the original competent authority.

2.5 Part C: Placing on the market products containing genetically modified organisms

2.5.1 General provisions

Part C of this directive stars with Article 10, which gives in paragraph 1, a set of general conditions before any product can be placed on the market.

These conditions are that:

- consent has been given under part B of the directive, meaning that no GMO can be placed on the market without a proper R & D stage;

- the product should comply with this directive and relevant product legislation.

Paragraph 2 of Article 10 indicates that the procedure for placing a product on the market shall not apply to products covered by Community legislation which includes a specific environmental risk assessment similar to that provided in this directive. The background of this provision is that:

- it is desirable to have only one procedure for placing products on the market;

- product legislation already contains procedures for placing on the market.

2.5.2 System

The same system of part B is found in part C.

Article 11, paragraph 1:

"before a GMO or a combination of GMOs are placed on the market as or in a product, the manufacturer or the importer to the Community shall submit a notification to the competent authority of the Member State where they are placed on the market for the first time".

Article 11, paragraph 5:

"the notifier may only proceed when he has received a written consent".

2.5.3 The notification

The first paragraph of Article 11 says that the notification shall include the information of Annex II information obtained from R & D releases and specific product information laid down in Annex III (use, labelling, packaging etc).

The final paragraph of Article 11 gives the steps to be taken when new information becomes available with regard to the risk of the product, analogous to Article 6.

2.5.4 Authorisation

The authorisation procedure of placing on the market is in fact a two step procedure. The first step is given by Article 12:

Paragraph 1: "On receipt and after acknowledgement of the notification the competent authority shall examine the conformity of the notification with the requirements of this directive".

Paragraph 2: "The competent authority shall respond in, within 90 days, by either:

- forwarding the dossier to the Commission with a favourable opinion;

- informing the notifier that the release does not fulfil the conditions of this directive and the notification is therefore rejected;

For calculating the waiting period of 90 days the period needed for the notifier to supply further information shall not be taken into account.

2.5.5 International consultation

The second step of the procedure is laid down in Article 13:

"The Commission shall immediately forward the dossier to the other Member States which may, within 60 days, present observations that are received from the other Member States".

When an objection is received and the competent authorities concerned cannot reach an agreement within these 60 days, the commission shall take a decision in accordance to a specific procedure.

2.5.6 Placing on the market: Community wide

One of the key articles of this part C is Article 15:

"A Member State may not restrict or impede, on grounds relating to the notification and written consent of a release under this directive, the placing on the market of product containing or consisting of GMOs which comply with the requirements of this directive".

This means that when a product has received a consent after the procedure of part C, no Member State may restrict the placing on the market on grounds of protecting human health or the environment.

When a Member State has justifiable reasons (eg new information) that such a product constitutes a risk, it may provisionally restrict the product, after which the Commission shall take a decision in accordance to a specific procedure (Article 16).

The commission shall publish a list of products which received consent under this directive (Article 17).

2.6 Part D: Final provisions

2.6.1 Confidentiality

Article 19:

"The Commission and the competent authorities shall not divulge to third parties any confidential information notified or exchanged under this directive and shall protect the intellectual property rights relating to the data received".

The notifier indicates what information he wants to be kept confidential, though be it that certain information cannot be kept confidential, like the name and address of the notifier and a description of the GMO, methods for monitoring and the evaluation of foreseeable effects.

It is the competent authority which decides, after consultation with the notifier, which information shall be kept confidential.

2.6.2 Commission procedure

The specific procedure mentioned before is explained in Article 21, which in general terms says that the Commission will be assisted by a Committee which votes by qualified majority. If measures envisaged by the Commission are not in accordance with the opinion of the committee, it will be submitted to the Council.

Appendix 2

Regulatory Position of Chymosin Produced from Genetically Engineered Micro-organisms

Introduction

From research stage to final product chymosin is subjected to three different regulatory regimes:

- Research or Small Scale Guidelines;

- Large Scale Guidelines and/or Regulations;

- the relevant Product Regulations.

The introduction of recombinant DNA technology has not led to a principal change in that scheme, although there was, and is, broad consensus that existing regulations should be adapted to accommodate this new technology.

EC directives from which the regulations arose are discussed in Appendix 1.

Focusing on the development of chymosin, a straightforward example of the application of recombinant DNA technology in a traditional fermentation process, some interesting aspects and consequences of this philosophy are discussed in this appendix.

Research

In most countries research involving recombinant micro-organisms is subjected to guidelines including a notification or authorisation procedure depending on the classification of the micro-organism. Until recently industry has adhered to these guidelines on a voluntary basis, however, these guidelines are now being incorporated in official regulation. Although there are national differences the general principles used are similar in each country.

The chymosin-producing organism *Kluyveromyces lactis* has been classified in The Netherlands in the lowest-risk category VMT (ie only Safe Microbiology Techniques need be used). In the UK this would be regarded as Group 1. This classification was an important first step in obtaining production approval and eventually product approval.

Production

The actual production of chymosin, both on a pilot plant and industrial scale were to take place in Gist-brocades' enzyme production plant in France.

Initially the national French Advisory Committee on Genetic Engineering was consulted, which resulted in the classification of the recombinant *K. lactis* at the lowest risk level L_1B_1. (Note how confusing the categorisation nomenclature used can be).

At a later stage French legislation regulating processing and fermentation was amended in such a way that production processes involving recombinant micro-organisms were included. According to this legislation, the production organism has been classified in the French equivalent of the OECD - GILSP level, allowing large scale production in existing production facilities. GILSP which stands for Good Industrial Large Scale Production criteria are described in the OECD report Recombinant DNA Safety Considerations.

They include the absence of pathogenicity and require an extended history of safe industrial use of the host organism.

Product

Food and Food Additive regulations tend to be very complicated and, especially when traditional foods are concerned, often overlapping.

A short overview of relevant regulations and guidelines both at a national and supranational level will be given. Following that, the requirements laid down in those regulations will be discussed including the way these various and varying requirements have been met.

Regulatory situation

National level

'Food Law' usually describes the basic principles that should be met before a food can be placed on the market. These principles most often encompass the following aspects:

- the food must not be dangerous to public health;

- the food must not impede fair trade;

- the food shall not mislead the consumer.

vertical legislation

Most of the time Food Law also contains what is called vertical legislation. This type of legislation describes often painstakingly how specific foods shall be prepared, thereby providing a certain protection against falsification of the use of lower quality ingredients. A negative aspect of this type of legislation is that it may cause serious trade barriers. As cheese is a very important traditional food item, vertical legislation in this field is well developed and directly affects the regulatory position of chymosin, used for curdling milk.

horizontal legislation

Food Law often also contains so-called horizontal legislation, which usually covers groups of additives with a common function eg preservatives, colours, emulsifiers etc. Although chymosin is not generally seen as an additive, it is sometimes included in horizontal legislation.

Apart from the types of legislation described above the industry sometimes creates its own set of regulations. This is mainly done to guarantee product quality to a higher degree than is possible with existing food legislation. Examples are the French Appellation d'Origine regulation for wine and cheese, equivalent Italian regulations and the German Reinheitsgebot for beer.

It is not yet clear what the consequences of the Completed Common Market in 1992 will be for the various types of legislation mentioned above.

It can safely be assumed that global harmonisation will be reached in the end, although there will always remain need and opportunities for industrial guidelines guaranteeing and protecting quality and origin of a specific product.

Supranational level

directives and guidelines

Here again various types of regulations can be distinguished. Basic differentiation is between legislation (eg EC directives) issued by a body with legislative power (EC) and guidelines issued by international organisations. These organisations are know by a series of acronyms (eg FAO/WHO - JECFA, AMFEP, OECD, IDF). We will meet some of these in more detail later. It should be noted however that the EC also issues guidelines.

Legislation

Although at a Community level a number of directives exist which regulate the use of food additives there is no community regulation governing the use of milk clotting enzymes. There is however a draft directive on Novel Foods being circulated which may eventually have consequences for the regulatory position of chymosin, although at this moment this is still uncertain.

Guidelines

Several international organisations have issued guidelines concerning food additives and enzymes. Most frequently these guidelines are formulated as specifications with the special aim of guaranteeing the safety of the preparation. The Joint Expert Committee on Food Additives (JECFA) or FAO (UN Food and Agriculture Organisation) and WHO (World Health Organisation) has issued general microbiological and chemical specifications of identity and purity of enzyme preparations.

These specifications are widely accepted by regulatory authorities all over the world and are considered to be of fundamental importance in the safety evaluation process of enzyme preparations.

The Association of Microbial Food Enzyme Producers (AMFEP) has also issued specifications which basically are the same as the JECFA specifications. Moreover, this Association has also issued general guidelines for the compilation of a safety dossier for enzyme preparations. As mentioned before, the OECD and the EC have formulated criteria for the classification of genetically modified (micro-)organisms (see Appendix 1). Although not directly related to the authorisation of chymosin itself it will be evident that the classification of the production organism *Kluyveromyces lactis* as a safe organism is of prime importance. In the OECD terminology a safe organism is characterised as having a Good Industrial Large Scale Practice (GILSP) status.

Finally the International Dairy Federation (IDF) has formulated guidelines for rennet preparations. In order to obtain acceptance from the dairy industry - the sole customer for chymosin:- it is necessary to comply with these specifications. A working group of the IDF is in the process of formulating an official IDF position on enzymes produced with recombinant micro-organisms.

The registration procedure

identification of target markets

After having defined the target market countries for chymosin, enquiries were started into regulatory requirements for the authorisation of chymosin in those countries. Obviously target countries were selected on the basis of the importance of their cheese industry. Complexity of the registration procedure has not been a decisive factor in that selection process. Enquiries were made directly with relevant national authorities, but also through industrial organisations, universities and independent institutes. This was thought to be necessary because of the innovative character of the product and the fact that the food industry was unacquainted with it.

product safety

After this round of discussions and after careful examination of existing national regulations in the dairy and enzyme field it was decided to develop a "two legged" approach concerning the safety aspects along the following lines:

- the safety of the product should be demonstrated through an extensive toxicological testing programme and a detailed characterisation of the production strain;

- the similarity of chymosin to the chymosin from calf rennet and its increased purity compared with the latter should be proved through an extensive chemico-physical and microbiological testing program, thereby adding further evidence to the safety of the product.

product efficacy

A further outcome of the consultations was that it was absolutely necessary to have firm proof of the efficacy of the product in actual cheese making, preferably at a national level because of the wide variety of types of cheese and cheese making processes. Although not always described and defined as a requirement for obtaining authorisation this is vital in order to get the industrial support to prove to authorities that obviously a need for the product existed.

To that purpose an extensive trial programme was initiated in the target countries, predominantly at independent dairy institutes or universities.

Safety programme

In principle, the fact that the production organism has been genetically manipulated should not have any influence on the safety evaluation as such.

Toxicological tests are equally applicable to products made by conventional methods as to those made by new methods. Their predictive value is the same in both cases.

At the time of chymosin development there was not a single instance where regulations existed relating to the application of recombinant DNA technology in the development of a production organism. Nevertheless it was deemed necessary to address some basic issues preceding the safety assessment of the product.

As a first step the safety of the production organism had to be proven. Criteria for these are:

- absence of pathogenicity;

- stability of inserted recombinant DNA;

- characterisation of inserted DNA and limitation of that DNA to those sequences needed for expression and, if applicable excretion. These preliminary safety measures are mentioned by the US (FDA) in a guideline called "Points to Consider" issued in 1985.

If production and recovery processes were such that no living recombinant organism was present in the final product no further measures were necessary to reduce the ability of the organism to survive in the environment (including within the product). It was highly desirable to avoid the use of any markers that code for the resistance of antibiotics that are currently used in human or veterinary medicine to avoid further spreading of resistance to those antibiotics. If these markers had been used precautions would have been needed to prevent DNA carrying those genes from being present in a transformable for in the final product.

Together with the aspects mentioned earlier, this has resulted in the production of the following safety dossier:

Part I Production organism	Questions tackled
Pathogenicity	Does it cause disease (test in animal, history)?
Stability of DNA	Is the inserted DNA stable?
Characterisation of DNA	Can all of the inserted DNA be accounted for and can samples be easily compared?
Part II Product Specifications	
Physico-chemical characterisation	Can the product be described in defined physical terms?
Microbiological specifications	Is it free of contaminant micro-organisms or meet particular criteria?
Chemical specifications	Can the product be defined in terms of a number of chemical criteria?
Part III Product Toxicology	
Short term studies	Does a large dose of the product cause damage?
Subchronic study	Do small doses given over a long period prove to be toxic?
Mutagenicity studies	Does the product cause mutation? (AMES test)
Allergenicity study	Does the product cause allergic reactions in recipients?
Part IV Product Efficacy	How active is the product? How does it compare with existing products?
Target cheese types	Which cheese types will be produced using the new product?

All in all, this package has proven to be acceptable to authorities in every country where Gist-brocades has filed a petition. This includes a notice of filing as a GRAS (Generally Regarded As Safe) substance by the USA FDA and formal clearance so far in Switzerland, UK and France. Toxicological clearance has been given in Germany.

Appendix 3

Units of measurement

For historical reasons a number of different units of measurement have evolved. The literature reflects these different systems. In the 1960s many international scientific bodies recommended the standardisation of names and symbols and a universally accepted set of units. These units, SI units (Systeme Internationale de Unites) were based on the definition of: metre (m), kilogram (kg); second (s); ampare (A); mole (mol) and candela (cd). Although, in the intervening period, these units have been widely adopted, their adoption has not been universal. This is especially true in the biological sciences.

It is, therefore, necessary to know both the SI units and the older systems and to be able to interconvert between both sets.

The BIOTOL series of texts predominantly uses SI units. However, in areas of activity where their use is not common, other units have been used. Tables 1 and 2 below provides some alternative methods of expressing various physical quantities. Table 3 provides prefixes which are commonly used.

Mass (S1 unit: kg)	Length (S1 unit: m)	Volume (S1 unit: m^3)	Energy (S1 unit: $J = kg\ m^2\ s^{-2}$)
$g = 10^{-3}\,kg$	$cm = 10^{-2}\,m$	$l = dm^3 = 10^{-3}\,m^3$	$cal = 4.184\ J$
$mg = 10^{-3}\,g = 10^{-6}\,kg$	$Å = 10^{-10}\,m$	$dl = 100\ ml = 100\ cm^3$	$erg = 10^{-7}\ J$
$\mu g = 10^{-6}\,g = 10^{-9}\,kg$	$nm = 10^{-9}\,m = 10Å$	$ml = cm^3 = 10^{-6}\,m^3$	$eV = 1.602 \times 10^{-19}\ J$
	$pm = 10^{-12}\,m = 10^{-2}\,Å$	$\mu l = 10^{-3}\,cm^3$	

Table 1 Units for physical quantities

Concentration (SI units: mol m^{-3})

a) $M = mol\ l^{-1} = mol\ dm^{-3} = 10^3\ mol\ m^{-3}$

b) $mg\ l^{-1} = \mu g\ cm^{-3} = ppm = 10^{-3}\ g\ dm^{-3}$

c) $\mu g\ g^{-1} = ppm = 10^{-6}\ g\ g^{-1}$

d) $ng\ cm^{-3} = 10^{-6}\ g\ dm^{-3}$

e) $ng\ dm^{-3} = pg\ cm^{-3}$

f) $pg\ g^{-1} = ppb = 10^{-12}\ g\ g^{-1}$

g) $mg\% = 10^{-2}\ g\ dm^{-3}$

h) $\mu g\% = 10^{-5}\ g\ dm^{-3}$

Table 2 Units for concentration

Fraction	Prefix	Symbol	Multiple	Prefix	Symbol
10^{-1}	deci	d	10	deka	da
10^{-2}	centi	c	10^2	hecto	h
10^{-3}	milli	m	10^3	kilo	k
10^{-6}	micro	μ	10^6	mega	M
10^{-9}	nano	n	10^9	giga	G
10^{-12}	pico	p	10^{12}	tera	T
10^{-15}	femto	f	10^{15}	peta	P
10^{-18}	atto	a	10^{18}	exa	E

Table 3 Prefixes for S1 units

Appendix 4

Chemical Nomenclature

Chemical nomenclature is quite a difficult issue especially in dealing with the complex chemicals of biological systems. To rigidly adhere to a strict systematic naming of compounds such as that of the International Union of Pure and Applied Chemistry (IUPAC) would lead to a cumbersome and overly complex text. BIOTOL has adopted a pragmatic approach by predominantly using the names or acronyms of chemicals most widely used in biologically-based activities. It is recognised however that there remains some potential for confusion amongst readers of different background. For example the simple structure CH_3COOH can be described as ethanoic acid or acetic acid depending on the environment or industry in which the compound is produced or used. To reduce such confusion, the BIOTOL series makes every effort to provide synonyms for compounds when they are first mentioned and to provide chemical structures where clarity and context demand.

Appendix 5

Abbreviations used for the common amino acids

Amino acid	Three-letter abbreviation	One-letter symbol
Alanine	Ala	A
Arginine	Arg	R
Asparagine	Asn	N
Aspartic acid	Asp	D
Asparagine or aspartic acid	Asx	B
Cysteine	Cys	C
Glutamine	Gln	Q
Glutamic acid	Glu	E
Glutamine or glutamic acid	Glx	Z
Glycine	Gly	G
Histidine	His	H
Isoleucine	Ile	I
Leucine	Leu	L
Lsyine	Lys	K
Methionine	Met	M
Phenylalanine	Phe	F
Proline	Pro	P
Serine	Ser	S
Threonine	Thr	T
Tryptophan	Trp	W
Tyrosine	Tyr	Y
Valine	Val	V

Index

A

B